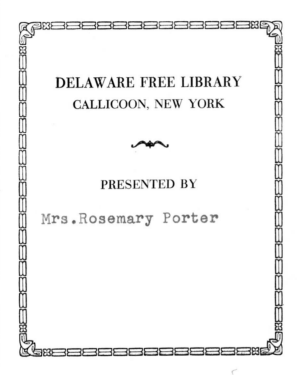

PARALLEL BOTANY

LEO LIONNI

PARALLEL BOTANY

Translated by

PATRICK CREAGH

ALFRED A. KNOPF   NEW YORK   1977

THIS IS A BORZOI BOOK
PUBLISHED BY ALFRED A. KNOPF, INC.
Copyright © 1977 by Leo Lionni
All rights reserved under International
and Pan-American Copyright Conventions.
Published in the United States
by Alfred A. Knopf, Inc., New York,
and simultaneously in Canada
by Random House of Canada Limited, Toronto.
Distributed by Random House, Inc., New York.

Grateful acknowledgment is made
to Macmillan Publishing Co., Inc.,
for permission to reprint one line from
"Poetry" from Collected Poems of Marianne Moore.
Copyright 1935 by Marianne Moore,
renewed 1963 by Marianne Moore and T. S. Eliot.

Library of Congress Cataloging in Publication Data
Lionni, Leo [date]
Parallel botany.
Translation of La botanica parallela.
1. Botany—Anecdotes, facetiae, satire, etc. I. Title.
PQ5984.L55B613   581'.0207   77–74985
ISBN 0–394–41055–6   0–394–73302–9 (pbk.)

Photographs are by Enzo Ragazzini

Manufactured in the United States of America
First American Edition

"Imaginary gardens with real toads in them."
—MARIANNE MOORE

CONTENTS

PLATES

PART ONE

INTRODUCTION

GENERAL INTRODUCTION

I n ancient times botany was part of a single science that included everything from medicine to the various skills of agriculture, and it was practiced by philosophers and barbers alike. At the famous medical school of Cos (fifth century B.C.) Hippocrates, and later Aristotle, laid the foundations of the scientific method. But it was Theocrastus, a pupil of Aristotle, who first worked out a rudimentary system for observing the vegetable world. The influence of his *Historia Plantarum* and *De Plantarum Causis* was passed on to later ages by Dioscorides, and is to be found lurking everywhere in the medieval herbaria composed by the scrivener monks in their cloistered gardens, with their humble little plants, each on its minute altar of earth, as still and perfect as waiting saints, wrapped in a solitude that defies time and the passing seasons.

After Gutenberg, plants also came to have a new iconography. Instead of the delicate washes applied with loving patience, and expressing the very essence of the petals and leaves, we now have the coarseness of woodcuts and the flat banality of printer's inks.

In 1560 Hieronymus Bock published a volume, illustrated with woodcuts, in which he described 567 of the 6,000 species of plant then known in the Western world, including, for the first time, tubers and mushrooms. "These," he wrote, "are not grasses, or roots, or flowers, or seeds, but simply the excess of humidity that is in the soil, in trees, in rotten wood and other putrescent things. It is from this dampness that all tubers and mushrooms spring. This we can tell from the fact that all mushrooms (and especially those used in our kitchens) most commonly grow when the weather is wet or stormy. The Ancients were in their time particularly struck by this and thought that tubers, not being born from seed, must in some way be connected with the sky. Porphyry himself expresses as much when he writes, 'Mushrooms and tubers are called the creatures of the Gods because they do not grow from seed like other living things.'"

Less than a century after the invention of printing, the conquistadores and the captains of the East India companies showered an astounded Europe with a perfumed cornucopia from the gardens and jungles which had until then slept beyond the oceans. Hastily, thousands of new plants had to be named and placed within a rudimentary and inefficient system of classification.

It was not until the first half of the eighteenth century that the Swedish botanist Linnaeus created a system of botanical classification that seemed to be definitive, a botanical register where all the plants of the earth, present and future, could be given a name, a degree, and a brief description. Linnaeus published his *Systema Naturae* and in 1753 introduced the double nomenclature giving each plant two Latin names, one for the genus and the other for the species. By now no fewer than 300,000 plant names compose one enormous random poem that records, commemorates, describes, exalts, and celebrates all that man has discovered of the world of plants.

All seemed ready for the emergence of the new science. Freed from their obsession with classification, botanists began to ask themselves how and why plants behave as they do. Chemistry, physics, and genetics provided new instruments of research, while classification gave way to etiology, the study of origins. Botany, called upon to establish a logical and causal relationship between the morphological structure and the vital functions of plants by experimental methods, became a modern science.

The future seemed securely mapped out: from the small to the smaller still, and so *ad infinitum*. It was thought that at that point, paradoxically enough, would occur the sudden fusion of knowledge that would explain everything in the universe.

But the triumphant and comforting prospect of a research program gradually but inevitably unraveling itself over the centuries was destined to be severely jolted by the news of the discovery of the first parallel plants, of an unknown vegetal kingdom which, being by nature arbitrary and unforeseeable, appeared—and still appears

—to challenge not only the most recently acquired biological knowledge but also the traditional structures of logic.

"These organisms," writes Franco Russoli, "whose physical being is sometimes flabby and sometimes porous, at other times osseous but fragile, breaking open to display huge colonies of seeds or bulbs which grow and ferment in the blind hope of some vital metamorphosis, that seem to struggle against a soft but impenetrable skin— these abnormal creatures with pointed or horny protuberances, or petticoats, skirts and fringes of fibrils and pistils, articulations that are sometimes mucous and sometimes cartilaginous, might well belong to one of the great families of jungle flora, ambiguous, savage, and fascinating in their monstrous way. But they do not belong to any species in nature, nor would the most expert grafting ever succeed in bringing them into existence."[1]

Fig. 1 A vegetable-lamb or *Barometz,* from a sixteenth-century woodcut

When we think that in 1330 Friar Odorico of Pordenone, with truly angelic devotion, described a plant which gave birth to no less than a lamb (Fig. 1), and that as late as the seventeenth century, on the threshold of the first real scientific experiments, Claude Duret also spoke of trees which produced animals,[2] we cannot wonder if the discovery of a botany unconstrained by any known laws of nature has given rise to descriptions that do not always treat the real character of the new plants with objective accuracy. As Romeo Tassinelli puts it: "What are we to say of plants that sink their roots, not into the familiar soil of our planet, but into an infinitely distant oneiric humus, feeding on ethereal juices not susceptible to measurement? The plants of this kingdom appear to be extraneous to the well-ordered play of natural selection and the survival of the species. They do not lend themselves to the surest and best-tried methods of experiment, and resist the most elementary kinds of direct observation. Their etiology, their very existentiality, can be assigned no place among the things of our planet. In short," he concludes, "we ought not to speak of a vegetable kingdom, but of a vegetable anarchy."[3]

It was clear that to find a place within the Linnaean classification for plants that were possible, or at best probable, but in any case totally foreign to our known reality, would present insurmountable difficulties. It was Franco Russoli who coined the phrase "parallel botany," at the same time giving a name and a definition to what might be a science in itself, or might simply represent, *in toto*, the organisms which are the object of inquiry. But it sometimes happens that words possess a wisdom greater than their semantic density. By means of its implications of unalterable "otherness," the word "parallel" freed the scientists from the nightmare of seeing the traditional classifications virtually destroyed, and along with them the very basis of modern scientific methodology. Insofar as Wolotov is right in observing that if one of two sciences is parallel then by definition the other must be also, we are of the opinion that the somewhat cloudy ambiguity of the word must be taken to refer to a realm outside the established boundaries of our knowledge. "Once aware of its parallelism," says Remo Gavazzi, "we are forced to change the focus of our observation, to create new paths for inquiry and maybe also new instruments of perception, if we are to understand a reality that might formerly have appeared hostile to us."[4]

Every discovery, however small, implies a redefinition of everything that we have so far comfortably accepted as the only possible yardstick of reality. Thus, the discovery of this unusual and disquieting botany was bound to upset the illusory consistency of our previous notions of reality and unreality. "So much so," writes Dulieu, "that it is from these very notions that its plants, mysteriously alienated from the events of growth and decay which struggle for the dominion of the biosphere, appear to draw their vital juices, and thereby emerge, perennially immune, outside the sphere of normal perceptions and the links and associations of the memory, in a fashion quite 'other,' ambiguous, perverse, and beyond our ken. We are unable to grasp it because of the long-consecrated notion of reality which clings so obstinately, like a twining and perhaps poisonous ivy, to our logic."

Jacques Dulieu, director of the Biological Studies Center at Bovences and editor of the journal *Pensée,* owes his international reputation not only to his celebrated experiments into the vibratory and echoic language of the organisms living on the seabed, but also to his detailed and original critical analysis of Descartes. It may have been the fact that he was both a biologist and a philosopher that first led him to take an intense and serious interest in the new botany.

Criticizing the ideas that since the Enlightenment had been held to be the sure foundations of all our work in the sciences, in a historic interview for Radiodiffusion Française Dulieu recounted the strange events which led up to his intellectual crisis, to his controversial reevaluation of all ancient meanings, and to the formulation of new methods of research for the study of phenomena which "official" science refused to recognize as really existent.

His dramatic testimony was meant as a reply to those in French intellectual circles who could not understand how a biologist of his stature might, with such outspoken determination, have taken the risk of exploring new and seemingly esoteric trajectories, so full of snares and inevitable pitfalls, when his reputation as a scholar of exceptional flair and prudence seemed already to have assured him a place among the luminaries of science.

In his radio interview Dulieu told how, shortly after the end of the war, he was working in the botanical biology laboratory at the University of Hannanpur in Bengal. There he met Hamished Baribhai, famous for his studies not only in medical botany but also in Sanskrit literature, and particularly the Vedic texts. When they

met, Baribhai had just turned ninety-one years of age, but in mental
and physical agility he could still with ease match the young French
scholar, who at that time was one of the up-and-coming talents at
the Sorbonne. The two of them used to meet often in an "ashram"
on a hill, near the great temple dedicated to the monkey-god Hanuman.[5]

"One late afternoon, in the first glow of a long sunset, when the
city was veiled in a reddish smog and the acrid stench of burnt dung
rose even to the hilltop, Hamished Baribhai said to me, 'You are
always talking about the real and the unreal. If you promise to keep
it to yourself I will show you a new experiment. Come with me.' We
walked for half an hour toward the River Amshipat until we came
to the edge of a wood of gensum trees. There we came upon a
freshly whitewashed mud hut. The door was padlocked. Baribhai
took a bunch of keys out of his pocket and opened the door. 'There is
your reality,' he said with an ironic smile. I was rather dismayed by
what I saw. In the semidarkness inside the hut were two large white
gibbons. One was stretched out on a pile of straw, and appeared to
be dead. Even when we entered it did not move. Meanwhile the
other, without stirring from its place, began to rock nervously on its
paws, showing its teeth and emitting little shrill cries. 'Is that one
dead?' I asked, pointing to the other monkey, which had still not
shown the least sign of life. 'If that one is dead, so is the other,' was
Baribhai's answer. Then he added, spelling the words out slowly,
'You are looking at a single monkey.' Being perfectly accustomed to
the old man's witticisms I did not react to this absurd statement.
'What do you think they're doing, those two?' I asked, intending to
tease him. But Baribhai had already left the hut. I followed, wonder-
ing what on earth he was up to. Though the monkeys were secured
on long chains, I shut the door carefully behind me.

"Next to the hut there was a long, narrow vegetable garden, no
larger than a bowling alley,* completely surrounded by six-foot wire
netting topped with barbed wire. It made me think, involuntarily, of
a concentration camp for dwarfs. Inside the garden there were
three rows of plants, all fifty centimeters high and all exactly the
same. At first sight they looked like tomato plants, but the leaves
were very regular and rather swollen-looking, like those of certain

----

* Dulieu actually referred to an alley for playing *boules*, scarcely more than half the
length of an American bowling alley. [Translator's note.]

succulents. Baribhai took out his keys again and opened the gate. He went in, picked three leaves from one of the plants with meticulous care, then came out, closed the gate, snapped the padlock shut, and showed me the leaves. 'Do you want to see reality? Come with me and watch carefully.' We went back into the hut. The monkey which had been lying down had not moved at all, but at the sight of the leaves the other became extremely excited. I was a little scared, without really knowing why, and kept close to the door. Baribhai held out the leaves to the monkey, who tore them from his grasp with a lightning movement, then sat down and leaned against the wall like a Mexican peon, munching the leaves with obvious relish. But as it ate, its frantic gestures slowed down, the eyes which had followed our every movement with such lively interest began to close, and when it had finished the third leaf it slid down onto the ground and lay there on its back, as if it had fainted. But at the instant it fell, completely inert, the other monkey appeared to shudder. It opened its eyes, emitted a long groan, rose to its feet, and looked around aggressively and with suspicion. At first I failed to grasp what was going on, but then I suddenly remembered what Baribhai had said ('You are looking at a single monkey'). 'There is your reality,' said the old scientist for the third time. 'Let's go.' "

The radio interviewer was unable to disguise his incredulity, and Dulieu went on: "I could scarcely keep my legs under me. We left the hut. Baribhai closed the door and locked it. I confess I had to sit down on one of the two crates which I found by the wall of the hut. Baribhai sat on the other, and for a while the only sound was the occasional rattle of a chain. 'What does it mean?' I asked him at last, almost in a whisper. 'Let's be off,' said Baribhai, as if he hadn't heard the question. 'Let's go before it gets dark.' We walked toward the ashram. The sky was now a fiery red, and here and there on the plain below us the first lamps were already lit. Then Baribhai began to speak.

" 'My young friend,' he said, 'you ask me what it means. Well, if I could tell you I would be Krishna, Shiva, and Vishnu all rolled into one. Ten years ago I was at Damshapur, in Orissa State, and a colleague of mine there told me about the strange properties of a certain plant, *Antola enigmatica*,[6] which grows on the slopes of PL. II Mount Tanduba. The shepherds who graze their flocks of black goats in the region pick the leaves of this plant and chew them. One day I asked one of them why he chewed the leaves, and he replied,

"Because when I close my eyes I seem to have become a mirror, and in the mirror I see myself, backwards." So then I tried the leaves, and after a few minutes I saw myself sitting in front of me, like an old friend who had come to visit me. From subsequent experiments I found that the leaves of the *Antola* contain a substance comparable to mescaline, called metexcaline H.B. I grew the plants in the garden of my laboratory, experimenting with grafts from other hallucinogenic plants such as *Kalipta onirica,* and after many attempts I succeeded in increasing and varying the psychedelic properties of the leaves. The plants which you saw in the garden by the hut represent ten seasons of experimental grafting, ten years of research, and I have now managed to produce a form of hallucination which I call "paragemination." It manifests itself as the feeling, and indeed the certainty, that one's body has divided into two identical bodies, while the consciousness remains whole, and comparatively unchanged. A few months ago I tried it myself, and was so terrified that I decided in future to experiment only on monkeys. The subject becomes two bodies with a single consciousness that moves, according to particular circumstances, from one to the other. When one body is "inhabited" by the consciousness, the other remains inert and apparently lifeless. But the extraordinary and disturbing thing about it is not the hallucination, weird as it is, but the fact that it is perceptible by others. Hypothetical explanations are infinite, and faced with a phenomenon so novel and bizarre they all seem valid enough. Maybe the leaves eaten by the monkey emit secondary hallucinogenic effects within the surrounding area, so that we too are involved. In that case the inert form might be an illusion on our part. Maybe we are, in certain conditions, the victims of the monkey's hallucination, rather as according to the *Daharna* all living beings are characters in a dream of Lord Krishna's. And who knows, maybe the phenomenon ought to be viewed within our habitual reality, as some new and totally unexpected combination of experiences. Ultimately,' the old man added almost to himself, 'paragemination in itself is a rather banal phenomenon. The important thing is to experiment in order to discover the existence of new and tangible categories of reality.' "

Dulieu's testimony may appear irrelevant, disproportionate, and perhaps outside the scope of these essays. I have quoted it at length

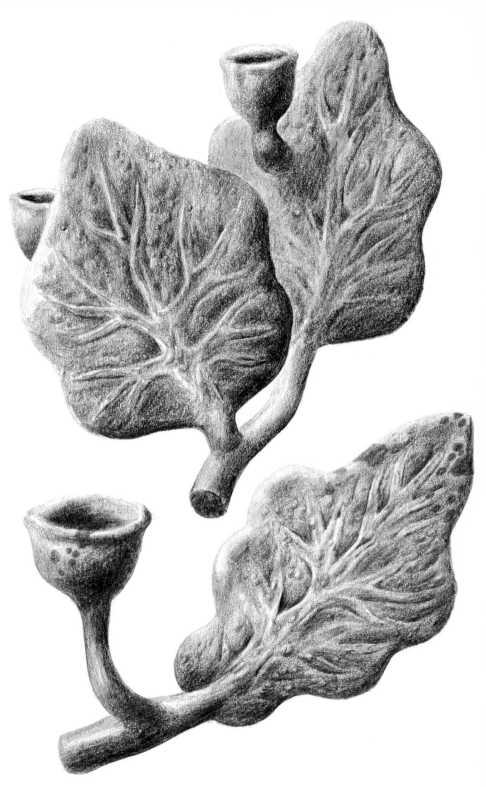

PL. II Leaves of *Antola enigmatica*

because I think that it indicates, though obliquely, something of the possibilities of our escaping from the age-old contradictions of logic; and above all because the great French biologist, always courageously open to new experiences, has since devoted himself almost exclusively to the study of parallel botany, contributing in a decisive way to defining the theoretical basis of the new science. In his book *Un autre jardin*⁷ Dulieu first asks himself the question: What is it that distinguishes the parallel plants from the supposedly real plants of normal botany?

For him there are clearly two levels, or perhaps even two types, of what is real, one on this side and one on the other side of the hedge. "On this side," he writes, "in our everyday garden, grow the rosemary, juniper, ferns and plane trees, perfectly tangible and visible. For these plants that have an illusory relationship with us, which in no way alters their existentiality, we are merely an event, an accident, and our presence, which to us seems so solid, laden with gravity, is to them no more than a momentary void in motion through the air. Reality is a quality that belongs to them, and we can exercise no rights over it.

"On the other side of the hedge, however, reality is ours. It is the absolute condition of all existence. The plants that grow there are real because we want them to be. If we find them intact in our memories, the same as when we saw them before, it is because we have invested them with the image that we have of them, with the opaque skin of our own confirmation. The plants that grow in that garden are not more or less real than those others which bend and sway in the wind of reason. Their reality, given them by us, is quite simply another and different reality."

That the parallel plants exist in the context of a reality that is certainly not that of "every day" is evident at first sight. Though from a distance their striking "plantness" may deceive us into imagining that we are concerned with one of the many freaks of our flora, we soon realize that the plants before our eyes must in fact belong to another realm entirely. Motionless, imperishable, isolated in an imaginary void, they seem to throw out a challenge to the ecological vortex that surrounds them. What chiefly strikes us about them is the absence of any tangible, familiar substance. This "matterlessness" of the parallel plants is a phenomenon peculiar to them, and is perhaps the thing which mainly distinguishes them from the ordinary plants around them.

The term "matterlessness," coined by Koolemans and widely used by both Dulieu and Fürhaus, may not be a very happy one, suggesting as it does the idea of invisibility, which except for certain abnormal situations is not generally true of parallel botany. "Para-materiality" would perhaps be a more correct word to describe the corporeality of plants that are usually characterized by a fairly solid presence, sometimes almost brutally intrusive, which makes them objectively perceptible to the same degree as all the other things in nature, even if their substance eludes chemical analysis and flouts all known laws of physics.

But "matterlessness" does suggest that apparent absence of veri-fiable structure on a cellular and molecular level common to all the parallel plants. Each individual species has some special anomaly of its own, of course, and these are more difficult to define and often far more disconcerting, though they are always attributable to some abnormal substance that rejects the most basic gravitational restric-tions. There are some plants, for instance, that appear clearly in photographs but are imperceptible to the naked eye. Some violate the normal rules of perspective, looking the same size however close or far they may be from us. Others are colorless, but under certain conditions reveal a profusion of colors of exceptional beauty. One of them has leaves with such a tangled maze of veins that it caused the extinction of a voracious insect that at one time had threatened the vegetation of an entire continent.

The parallel plants fall into two groups, but the distinction does not signify different evolutionary levels, as is the case with normal plants, which are divided into higher and lower orders. On the contrary, the two categories assigned to parallel plants are derived from the two ways in which the plants are perceived by us. Those of the first group are directly discernible by the senses and indirectly by instruments, while those of the second, far more mysterious and elusive, come to our knowledge only indirectly, through images, words, or other symbolic signs. The first group is certainly the larger, and contains the more widespread species. As Dulieu ob-serves, its plants are "the more parallel." Motionless in time ever since the strange mutation which triggered their metamorphosis, they have shared—some of them for millennia—the rather shabby history of the real world. But while all around them other plants grow, multiply, and disintegrate into humus, the parallel plants preserve their formal identity intact, like graven images.

If we are now in a position to perceive them, if we are able to observe, measure, and study them, it is in spite of the disturbing absence of any recognizable substance. This "matterlessness," referred to above, would seem to be the result of a sudden halt in time which for causes as yet unknown appears to have affected certain species of plant at various stages in the history of the vegetable kingdom.

Whereas other plants, now extinct, have disintegrated and left no further witness of their life on earth than the occasional fossil imprint or fragment of petrified bark, the parallel plants are, in Spinder's words, "fossils in themselves."[8] Neither dead nor alive—conditions both of which would imply a normal passage of time—they are still themselves, entire and perfect in their illusory corporeality after millennia of immobility. It is as if they had been suddenly torn out of time, emptied of matter and meaning, and given over to another order of existence. Like a memory that has taken on actuality, they have preserved of themselves only the outer appearance, a visible three-dimensionality without any substance. Most of these plants, though impervious to any violent acts of nature, disintegrate at the least contact with an object alien to their normal environment, dissolving into dust and leaving only a chemically inert white powder. Their behavior is similar to that of Egyptian mummies that have remained intact for thousands of years in their dark tombs, but which fall to pieces at the first ray of light, leaving only a spectral film of human substance in the bandages. Dulieu observes that these plants are in fact like mummified plants which a strange destiny has seen fit to immortalize not at the moment of death, but at the most significant moment of their life, to preserve in their undisturbed integrity, still protagonists of the landscape in which they stand, exuberant and happy.

The plants of the second group are also conditioned by abnormal and often incomprehensible temporal relationships. But instead of being permanently immersed in the constant flow of external time, they modulate their existence according to changing rhythms, which to our perceptions are unpredictable. While the plants of the first group are motionless *in time,* those of the second, chimeras of previous existences, move so to speak *outside of time,* in the man-made amorphous time of our own brains, in an unmeasurable series of sudden spurts and equally sudden halts in the past, in the future, and in the defunct present. They are the concrete image of this capricious non-time, parallel to the time which passes and in which we are accustomed to move.

This "parachronomy," as Spinder calls it, as opposed to the "chronostasis" of the other parallel plants, has implications which we have only recently begun to understand. It was Spinder himself, faced with phenomena that clearly overstepped the bounds of biology, who surmised that these plants can only be understood by means of the principles and methods of phenomenology and perhaps even of psycholinguistics. Connected to us by close psychosymbiotic links, their presence in a certain sense appears richer and "denser" than that of the plants of the first group, because they grow in the rhythm of our subjective time and eventually take the form of a long and intricate conceptual process. These plants, which for inexplicable reasons lost their real existentiality at some fairly remote point in real time, are today rediscoverable in the eventful landscape of our imaginations, where they reemerge from the authentic distant past, enriched with an ambiguous present, ready to be illustrated, described, and commented on.

"Parachrononomy" is therefore the key to their doubly parallel existence. Like the subjects of old portraits they are reborn today, after long repose in oblivion, with a double identity: the one which lives in our imaginations, and the other, now independent, which we see before us in its gilded frame, with its own reality.

In a paper read at the 1970 Antwerp Conference, Hermann Hoem stated: "All the things in the world dwell in us, in the mirror of our consciousness. All our gestures, even the most insignificant, are bound up in some way with a part of the world around us, altering the form of it and enriching it with new meanings. This applies also to our decision to divide the parallel plants into two groups. It reflects the coexistence of two important impulses in us: the impulse toward clarity and the impulse toward ambiguity. One might say that one group is the prose of parallel botany, while the other is the poetry. The plants of the first group are *subjected* to language *a posteriori;* those of the second are *born from* language, and verbal discourse is one of their preexistent conditions. Before being plants, they are words."

But it is in the nomenclature, perhaps because the names are naturally short, that these different relationships between plants and words are at their most convincing. The plant names of the first group reflect a sunny simplicity, as well as the particular circumstances of their origin and existence. Names such as "tiril" and "woodland sugartongs" are clearly descriptive, even though like all new words they are capable of giving rise to secondary images and associative ideas. "All names tell a story," says Hoem.

Names such as *"Solea"* and *"Giraluna"* actually precede the existence of the plants themselves and share, like a promise, in their very genesis. These names, which Jean Renon calls *"machines à faire poésie,"* are part of the substance of the plant, like a leaf, a stem, or a flower.

Although parallel botany appeared so suddenly and prominently upon the horizons of science, ten years passed before it was officially recognized. But it was little less than a miracle that in such a short time so much information and evidence could be collected and subjected to the necessary checks and counterchecks, and that contacts could be made on an international level between scientists and research workers, while specialized laboratories were set up in several countries. From the first sensational discovery of the woodland sugartongs in 1963 to the first Parallel Botany Conference in Antwerp in 1970 there was what Spinder has called a "parallel plant rush." News of fresh finds of plants and fossils, of legends and stories related to the subject poured in from all over the world, and there was scarcely an issue of any scientific journal without some theoretical article or bulletin of new discoveries. Books, doctoral theses, dissertations, and even new specialized journals piled up in the libraries of botanical and biological institutes, while in the laboratories work went ahead to improve or adapt the instruments to be used in documenting this new flora, so utterly strange, fragile and elusive. The Antwerp Conference, organized thanks to Cornelis Koolemans of the Royal University of Belgium, was in some sense intended to "place" the new science, to combine many individual efforts into one, to lay the theoretical basis for an understanding of the new phenomena, and if possible to arrive at some form of systematization, even though tentative and provisional.

Koolemans, who by a strange coincidence is the Go champion of Belgium, was in Japan for the finals of the Zendon Games[9] in Tokyo in the autumn of 1963. He had met Sugino Kinichi, a professor at Kyoto University and also a keen Go player, not long after the war at a conference on paleobotany in Paris. It was in fact

Sugino who on that occasion had introduced Koolemans to the game of Go, and without ever meeting again they had played interminable matches by correspondence. Koolemans tells how one of these intercontinental matches went on for sixteen months, and he estimates that between 1946 and 1963 their games of Go had cost the two biologists about twelve thousand dollars in postage, telephone calls and telegrams. When they finally met again in Tokyo in 1963, news PL. III came of the discovery of woodland tweezers in a wood near Owari, a find that was to have a dramatic impact on the biological sciences. Koolemans accompanied his friend on the first expedition and was so overwhelmed by the experience that he decided on the spot to devote himself entirely to the new botany. Although his work has been and still is chiefly in the organizational realm, Cornelis Koolemans is considered by his colleagues to be the first parallel botanist. Jacques Dulieu, in his closing speech at the Antwerp Conference, observed that if it had not been for the extraordinary intuition of the Belgian biologist, who from a single plant deduced the existence of a whole new vegetable kingdom, parallel botany would still remain undiscovered.

The idea of dividing the plants of the new botany into two groups was formally proposed at the conference by Koolemans himself and was accepted unanimously by the sixty-eight delegates after scarcely more than an hour of deliberations. But when it came to naming the two groups, things went rather differently: The debate lasted for nearly two days, but the lively and sometimes factious speeches did serve the purpose of better defining the differences between the two groups, which in the initial euphoria of the conference had not really been outlined with sufficient clarity. The names proposed by the various speakers, in fact, could not avoid describing the characteristics of the plants to which they referred, and thus what should have been made clear in the discussion of the first day's agenda ended up taking the form of a long debate on nomenclature.

The first proposal was made by Spinder. Max Spinder, a scientist of great intuition and inexhaustible energy, is Professor of Urban Botany at the University of Hemmungen. His is a new chair, established at his own insistence, for the study of plant life in urban areas. It may well have been his observation of urban plants, forced to survive in the most preposterous ecological conditions, that led the Swiss botanist to take an increasingly intense interest in parallel

PL. III Woodland tweezers at the base of a *ben* tree

botany. His laboratory, probably the best equipped in Europe, has provided him with the ideal conditions in which to carry out basic research into the new science. This research has been amply documented in his recent volume, *Parallelbotanik—Forschungen und Hypothesen,* published by Hansen Verlag of Zürich.

In his address to the conference, Spinder reminded his colleagues that in spite of a certain descriptive function, the name of the first group could in fact be entirely arbitrary, while that of the second group ought, like the names of its plants, to express the dreamlike quality, the vagueness, the profound ambiguity characteristic of them. At the same time, he said, it would be risky to burden the taxonomy of a science as young as parallel botany with a nomenclature which subsequent discoveries or experiments might prove ridiculous. "But in spite of this unresolvable dilemma," he concluded, "if we are to avoid cumbrous circumlocutions where any word or sign, even the most abstract, would really suffice to show what we are referring to, it is absolutely necessary that we come to

Fig. 2 Max Spinder

a decision." Realizing that his colleagues would certainly propose names that contained some allusion to the most salient qualities of the two groups, he himself suggested for the really existent, tangible, and visible plants the term "paraverophytes," while for the second group he suggested the name "anverophytes."

It was this latter suggestion that caused a debate which soon degenerated from the scientific and technical level into useless pseudophilosophical disquisitions on the nature of the real and the unreal, while semeiology, phenomenology, and even ethics were dragged in to support a variety of opinions.

One of the most interesting and significant speeches was that of Jacques Dulieu. To the admiration and amazement of the delegates, the French biologist quoted from memory the four pages of Descartes concerning the division of the things of the world into *res cogitans* and *res extensa,* and pointed out how the two groups to be named were a clear and perfect example of the Cartesian categories. He ended by suggesting the names "extendophytes" and "cogitandophytes."

We cannot here give all the proposals in full, varying as they did from the hagiographic "Spinderenses" and "Koolemanenses" to the clumsily allusive "parabiogenes" and "imagogenes," and from "heliophytes" and "selenophytes" to "oneirophytes" and "diodenophytes."

At the end of the second day of this absurd debate, Ezio Antinelli of the Centro Lombardo per le Scienze Applicate referred the delegates to an article he had written in *Vita Parallela,* the first periodical in this field intended for the general public, and repeated his suggestion that *all* plants, both common and parallel, should be divided into "existent" and "inductive." The "existent" plants, he said, revealed themselves as real through the evidence of the senses and scientific instruments. They in turn should be subdivided into "vital" (e.g. pinetree, carrot, narcissus) and "paravital" (e.g. tiril, *Plumosa, Labirintiana*). The "inductive" plants, on the other hand, are those which "live in a state of intention, waiting to take on form and solidity from an act of will on our part, which describes them." In other words, while recognizing two substantially different groups of plants in parallel botany, Antinelli wished to assign one of them, by using the ambiguous term "paravital," to the borderland of traditional botany, and to isolate the plants which he calls "inductive," and which he considers truly parallel, in a category of their own.

It was the Australian botanist Jonathan Hamston who reminded his colleagues of Spinder's warning, and by so doing brought them back to common sense. He begged the conference to avoid evocative or descriptive names, or those with too specific a content, and to leave the youthful science with enough elbowroom in the matter of terminology. He suggested calling the two groups of plants "Alpha" and "Beta" as a provisional solution. This was welcomed with visible relief by speakers and delegates alike, and on a motion proposed by Dulieu and seconded by Antinelli it was accepted unanimously.

ORIGINS

The most recent theories in the field of paleobotany trace the origins of the two botanies to aquatic protoplants, prechlorophyllic algae of the Ambrian era to which, unfortunately, we have very few clues, and those practically undecipherable. We do, however, possess fossil remains of the next phase of plant life, when a marine alga first put down roots on terra firma, thus becoming the matrix of all vegetation on dry land. These fossils were recently discovered in the Tiefenau Valley and its surrounding mountains by a group of German paleontologists led by Johann Fleckhaus. This tangible evidence appears once and for all to confirm the thesis which the paleontologist Gustav Morgentsen of Pålen University put forward at the 1942 European Conference on Botanic History at Smorsk. Those were the days when the Nazi armies were at the gates of Stalingrad, and the dramatic events of the war were destined to obscure the scientific importance of that speech, which, we must add, was received by many delegates with unconcealed skepticism.

However, the recent discovery of the Tiefenau fossils seems to have removed all doubts as to the validity of the hypothesis put forward by the celebrated Norwegian scientist. Twenty years after that historic event they are accepted by the scientific community as a basic dictum without which the explanation we can now give of the evolutionary "grand design" would be no more than a tentative sketch. In the scientific supplement issued by the *Smorskaya Gazeta* on the occasion of that memorable conference, Morgentsen wrote a brief popular account of his theory, which is now known as Morgentsen's theory of the great winds. He held that the origin of

Fig. 3 Gustav Morgentsen

plant life on dry land is to be assigned to the second half of the Ambrian era, when for causes still unknown to us the atmosphere was violently disturbed by vast hurricanes which circled the globe for thousands of years. The continents were then huge bare islands without the least sign of life, while in the oceans self-propelling multicellular organisms had already developed. There were large areas of floating algae at various depths. These plants were the first to utilize solar energy directly through the operation of a particular substance, chlorophyll, and by this means to transform water and carbon dioxide into the sugars and starches needed for their life process.

There were four types of these algae, three colored and one colorless. The colored types, modified structurally to adapt themselves to the increasing saltiness of the oceans, have survived down to our times. The best known are the green algae. Their color is derived from chlorophyll, which in the red and brown algae is disguised by pigments of other colors: phycoerythrin and phycoxantin. But the most common alga during the Talocene and Ambrian eras was the *Lepelara*, which has been extinct for at least 100 million years and which must be considered as the true parent of all plant life on dry land. The *Lepelara* was a single-celled alga shaped like a spoon (the name comes from the Dutch word for spoon, *lepel*), which on account of its low specific gravity floated nearer to the surface of the water than the other algae. It too achieved nourishment by photosynthesis, but through the medium of a colorless and autogenetic substance similar to chlorophyll, called atrophyll. This was present both in the nucleus, which was in the middle of the rounded and slightly swollen part of the cell, and in the rudimentary canal that ran down the "tail" or "handle" of the cell. The *Lepelara* was the oldest of the algae, and like many of the organisms then living in the seas it was completely transparent. As it was invisible, the exigencies of survival and even "self-presentation" did not demand that it have any particular size. There were *Lepelara* as big as oak trees, others as tiny as the frond of a maidenhair fern. Millions of these algae lay floating near the motionless surface of the waters.

But this primordial paradise, spread like an immense spangled embroidery beneath the monotonous succession of sun and moon, was one day touched by a sudden tremor. A breeze of unknown origin brushed it like the wing of a gull. Sporadic winds began to ripple the surface, and then to rouse it into waves. Scattered storms and waterspouts tore the algae from the water, hurling them back in chaotic frenzy one upon the other. Eventually a number of violent hurricanes came into collision, probably in the area where the Sargasso Sea is now, and this started the rotary movement which was destined for thousands of years to lash the seas and all that floated in them with insane and relentless fury. Whirled up in the spray of the shattered waves, the *Lepelara* were flung round and round the world, caught in an endless cyclone, to fall back into the raging seas, to disintegrate in the air, or to fall, alone or in groups, on the sterile soil of the continents and great islands. Then one day,

PL. IV

PL. IV Algal *Lepelara*

the fury of the hurricanes abated and calm returned to the earth. Millions of *Lepelara* of all sizes, piled up in crevices, against the rocky cliffs, between the boulders, and in every little crack or fold in the earth's surface, began slowly to die, still wet with the spray.

"But see," writes Morgentsen, "how one *Lepelara*, a 'guided case' in Teilhard de Chardin's phrase, with a sudden mutating burst of inexplicable invention, begins to breathe, to suck, to absorb oxygen, hydrogen, and minerals from the wet earth that partially covers it. Slowly the inert form begins to swell, to become, to be. A wash of color suffuses it, quite faint at first, then more and more intense, condensing to a strange opacity. The transparent alga is now alive and green, ready for the sign from destiny, the gesture that will tell it to rise and grow on dry land, the very first plant in all the earth." PL. V

The theory of the great winds was attacked by some of the leading paleontologists and biologists of the time with no lack of irony. Their doubts were perhaps exacerbated by the excessive simplification of the ideas of Morgentsen, and by the lyrical tone of the paper, which at that time was considered in bad taste at a scientific conference. But the younger delegates greeted it as a revelation. Among the Norwegian scientist's most enthusiastic supporters was Spinder, who had attended Morgentsen's courses at Pålen and was even then only thirty years of age. Building upon his teacher's ideas he developed his theory of the permanence of form, in which he attempted to show that all plants now extant are derived in some manner from the basic form of the *Lepelara*. According to this theory, exact analogies of outline with those of the original form, the *Urform*, provided confirmation that there was one single morphological scheme within which all the earth's flora gave evidence of its evolutionary link with the *Lepelara*. In support of this, Spinder wrote a book listing and comparing 128 varieties of plant, which he illustrates with meticulous realism in a series of drawings of such beauty that they alone would be a sufficient justification for the book. The theory was bold and original, but in spite of the ample documentation he provided it was not convincing and met with no more success than the work which had inspired it. Spinder himself recently rejected it as too arbitrary, and a mere "youthful caprice." But two years ago the Swiss scientist published a detailed study of the Tiefenau finds, culminating in a most meticulous reconstruction of the alga *Lepelara*, now recognized as the legitimate forebear of all plant life.

The Tiefenau fossils, which are provisionally displayed in a small room in the Hochstadt[1] Town Hall, are seven in number. Six of them are about twenty centimeters high while one, the so-called *Lepelara Morgentsenii,* is bigger, about seventy-two centimeters. Of the six smaller specimens only one bears the complete imprint of the alga, while two are unfortunately in such bad condition that the form can scarcely be recognized. The *L. Morgentsenii* is broken into three parts, but the imprint of the plant is complete except for one unimportant portion of the caudal section (corresponding to the handle of the "spoon"). It is a truly magnificent specimen, remarkable for the clarity of its outlines and its precision of detail. It was the analysis of this fossil that enabled Spinder to reconstruct the anatomy of the *Lepelara* in its most minute particulars. According to the biologist, the "protoplasm" of the *Lepelara,* which is its living substance, was contained in a rather thick and extremely tough portion of its anatomy. This membrane became much thinner toward the end of the "tail," where a plasmodesma with an exceptionally large opening enabled the cell to absorb oxygen, hydrogen, and other nutritive elements by osmosis. Later in its history the *Lepelara* developed its first rudimentary root system here.

Unlike the other algae the *Lepelara* had a proper nucleus, filled with a liquid called karyolymph, and here the filaments of chromatin wound themselves into a tangle of nucleoles, the latter also being composed of spiraling filaments pressed closely together.

We have learned from the most recent biological studies that the *Lepelara* must have contained in its DNA spiral not only its own plan for future development but the entire evolutionary program of plant life on earth. Piero Leonardi writes: "We are forced to think that these protoorganisms, in their basic makeup, had tendencies that were not left to the mercy of purely fortuitous circumstances, but were coordinated *ab initio* with a view to producing an organic and interdependent development of all living things, both vegetable and animal."[2] Seen in the light of this "law of guided complexities" (Teilhard de Chardin), according to which all living organisms are responsible for the development and equilibrium of the biosphere, the *Lepelara* takes on an importance which neither Morgentsen nor Spinder could possibly have imagined.

If the *Lepelara* may be considered the forebear of all plant life on the globe, the *Tirillus,* judging from what we may deduce from the

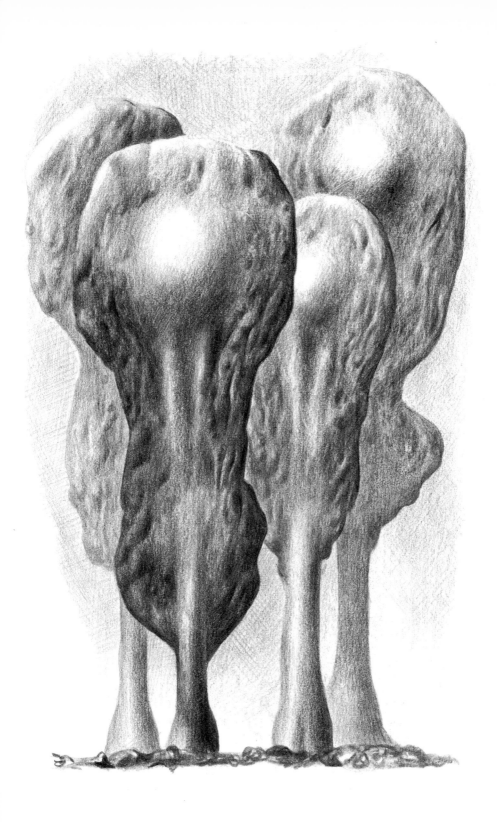

PL. V *Lepelara terrestris*

fossils discovered in various parts of the world, is almost certainly the first parallel plant.

The startling discovery of an extensive stratum of fossil *Tirillus* near Ham-el-Dour in the Luristan desert was made by the French

Fig. 4 Jeanne Hélène Bigny

Fig. 5 The clairvoyant Farah Apsalah Hamid

paleobotanist Jeanne Hélène Bigny, wife of the celebrated syndrolo-
gist Pierre-Paul Bigny, who for some years has been carrying out
important research at the Sorbonne, chiefly into the hydromagnetic
radiations of biomorphic fields. In the background of this discovery,
which after that of the Tiefenau *Lepelara* is without any doubt the
most important event in the botanical parapaleontology of the century, there is a curious interweaving of scientific zeal and personal
eccentricity that is worth describing here. The story, which in some
of its unusual facets involves parapsychology and psycholinguistics,
was reported by Roger Dadin in a recent issue of the women's
magazine *Nous*.

Jeanne Hélène Bigny, who like her husband teaches at the Sorbonne, is a scientist known for her personal eccentricities as well as
her important discoveries in paleontology, and while still assistant
to Marcel Declerque she achieved some notoriety. One day she gave
voice to an intuition that the fossilized remains of a large *Ankylo-
saurus* were to be found in the neighborhood of the Madeleine, in

PL. VI Fossil *Lepelara* from Tiefenau

the heart of Paris. The young paleontologist, whose uncle Jacob Charbin happened at the time to be a minister, made such a fuss that she was given permission to make a trial dig under the sidewalk that flanks the church, right opposite the Restaurant Duval. Madame Bigny did not find the fossil, but to the astonishment of all present, including Roger Dadin, then a reporter on the *Figaro de Paris*, she unearthed nothing less than the complete skeleton of a *Ceratopsius monoclonius*, which is now on view in the Dinosaur Room of the Musée Grignet.

From that time on, apart from the paleontology which was the field in which she specialized, Madame Bigny began to delve secretly into parapsychology. She began, occasionally at first, to frequent the famous Persian clairvoyant Farah Apsalah Hamid, who among her devotees could boast of such persons as Jean-Roland Bartand, Remi Antinos, Marcel Fouquet, and, so the rumor goes, even the president of the Chambre des Députés, Robert-Marie Autrac. But once Madame Bigny's interest in parallel botany had been stirred by her friend Gismonde Pascain, director of the Laboratory of the Jardin des Plantes, her visits to the beautiful Persian medium became more frequent.

It was on the fearful evening of August 14, 1971, remembered by Parisians for the violent storms that plunged all the *arrondissements* north of the Seine into total darkness and brought the entire Métro system to a standstill, that Jeanne Hélène Bigny, obsessed by strange presentiments, was sitting at the little round table opposite Madame Hamid. Flashes of lightning, filtered through curtains that flapped wildly in the half-open windows like torn shreds of sails, fell intermittently on the faces of the two women, causing them to float for an instant in the heavy darkness of the room. In spite of the continual rumble of the thunder, and apparently ignoring the frayed nerves of her client, the medium ceaselessly poured forth words, disconnected and incomprehensible. Her fingers, covered with gold rings laden with emeralds and amethysts, vaguely caressed the sinister object that stood with its sharp claws dug firmly into the thick red velvet tablecloth. It was a stuffed salamander whose crystal eyes, unnaturally large and protruding, blazed into life at every flash of lightning.

The scientist sought in vain for some logical connection between the clairvoyant's words, broken as they were by crashes of thunder, and in the end left the apartment in a state of anguish and confusion. It was some days later, in the quiet of her study in the Avenue des Ardennes, that those vague and disconnected phrases suddenly began to drift back into her mind, and the names of Ham-el-Dour, Sarab Bainah, and Tihir El emerged perfectly clear and precise. Madame Bigny had not the least idea what places or people these names might refer to, and yet they had come into her memory with all the solidity of things of our childhood which we have long forgotten and then happen to find in some dusty old trunk.

For days on end she searched for some explanation of the three names, which were clearly of Arabic or Persian origin. She appealed to Madame Hamid more than once, but the medium was unable to shed any light on their meaning; in fact, she denied ever having pronounced the names. But one day, to her amazement, Madame Bigny found them by chance in an old *Guide Bleu* to the Middle East.

Sarab Bainah turned out to be an area in the great desert zone of Ham-el-Dour in eastern Luristan, and Tihir El corresponded almost exactly to the name of a village in that area, the center of an oasis at the meeting point of the three great caravan routes that cross the desert. Research at the Institute of Middle Eastern Geology revealed that it was near this village that Iranian archaeologists, in collaboration with a team from the University of Pennsylvania, had discovered a necropolis formed of extremely deep burial shafts, in which the successive levels of the tombs indicated a historical continuity of nearly four millennia.

Madame Bigny set off for the desert, sure of having been chosen to make a sensational discovery. She reached the dig at Tihir El at the end of November, and was there able to study a number of very primitive artifacts that had been found at the bottom level of the shaft provisionally designated by the letter F. Among the finds were some fragments of limestone bearing imprints resembling the protocuneiform script of the famous "Gar Tablets," which had been found a few years earlier in the necropolis of Dum Gar Pachinah, only a few hundred kilometers from Tihir El. But while the archaeologists, struck by this surprising analogy, began to point out the significant connections between the two burial places, Madame Bigny at once recognized the clearly fossil origin of the fragments. It was the discovery of these fossils (Fig. 6), in fact, which prompted her to undertake the research that led to the discovery of the famous fossiliferous layer now known as the Bigny Layer.

Many hypotheses have been put forward to explain the mystery of why the name of the desert village of Tihir El so closely resembles that of the fossilized plants found underground in its vicinity. Roger Moseley went into the matter quite recently, and published the results of his research in the *Review of Psycholinguistics*. Moseley's main thesis is concerned with the unusual relationship between the parallel plant and its name, which is unique in the history of semeiotics because, as he says, it lacks one of the ele-

Fig. 6 Fossil tirils from the Bigny Layer

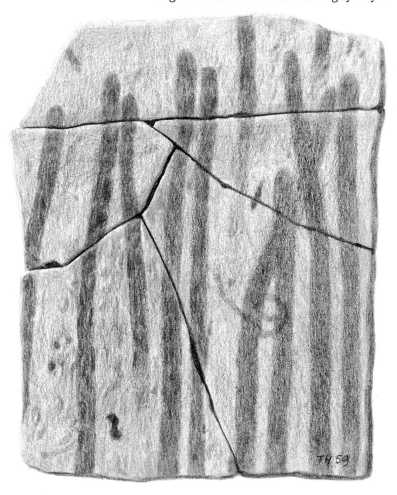

ments of the Bodenbach-Kordobsky triangle: name-thing-thing. Else-where we have seen how in certain parallel plants the name pre-ceeds the physical existence of the plant itself. According to Moseley, in the case of the tiril the name exists *independently* of the thing named, almost as if it were a reality in itself, with a substance of its own instead of a mere symbolic function—the very substance which the plant has been denied. Moseley calls this process "intui-tive codification," and as a case in point he cites the name of the village of Tihir El, founded at the time of Darius, when the Bigny Layer had for millions of years already lain at what was then an inaccessible depth.

Domenico Fantero, who has been responsible for a number of excavations in the area, makes the objection that the tiril was known of at the time of Darius in places not far from Tihir El, so that it is by no means out of the question that there were fields of tirils in the neighborhood of the village at the time it was founded. But Moseley quite justly points out that the tiril is never found superimposed on its own earlier beds: "For the tiril to replace its own dead would be an unimaginable compromise with time." He also observes that neither the name Tihir El nor its variant Ti-Hirel has any meaning in the languages and dialects of the largely nomadic peoples who have lived at various times in the Ham-el-Dour desert. Nor, says Moseley, can we suppose for one moment that the name commemorates some historical or divine personage, for in all local religious beliefs earlier than the age of Darius it was "forbidden to transfer the names of kings or gods to the common things of the earth."

With regard to this, Moseley draws attention to the onomato-clastic edict of Aktur,[3] which forbade the use of all proper names of persons or places except that of the Emperor himself. This edict resulted in such confusion that the administration of the Empire completely collapsed. Moseley, incidentally, went to Paris and closely questioned the medium Madame Hamid, who assured him that the name had simply "popped out of her mouth," that she could not have known of the existence of the village, and that apart from everything else she had never heard of the Sarab Bainah desert. Moseley later found out that Madame Hamid was not even Persian, but was born at Arles, in Provence, of a Basque father and a French mother. In her youth she had been on the stage, but with scant success. Moseley noticed that on the wall of her room she had a portrait of Sarah Bernhardt, and thus he struck upon the almost incredible similarity between the name of the great actress and that of the Luristan desert of Sarab Bainah. In this writings he often cites this as a typical example of intuitive codification.

In his article in the *Review of Psycholinguistics*, Moseley traces the history of the name *tiril* through its many transformations, pointing out a number of evolutionary hiatuses that suggest the existence of what he calls "word islands." He explains that in spite of the total absence of cultural links, for certain kinds of things these "islands" develop analogous terminologies, thus defying any kind of traceable etymological evolution. They tend to confirm the theory that the name existed earlier, and independent of any link with things or ideas. Among the projections of the word *tiril* in a number of word islands, Moseley quotes the extreme case of the Tabongo of the Mogo. Without having the least knowledge of the plant they use the expression *ti-ri-hil* as a sort of generalized utterance, an exclamation which has no reference to anything whatever, a perfectly abstract swearword.

The Bigny Layer is the most important evidence we have regarding parallel life in prehistory. The size of the bed has not yet been accurately assessed, but it quite possibly extends to three or four hectares. By an odd coincidence other fossil tirils were brought to light in various parts of the world only a few months after the discovery at Tihir El. Though of less importance than the Bigny Layer, they have nonetheless contributed to our knowledge of the plant, chiefly with regard to its very widespread geographical distribution and its survival under the most heterogeneous geological and climatic conditions. A very useful little book published by the Tirillus Society of America, *The Fossil Tirillus*, lists and describes all the sites where fossil remains of the plants have been found, examining their paleontological characteristics and listing the museums, institutes, and private collections in which the fossils are preserved.

While paleontology has provided fossil evidence of the origin of vegetation on the earth and of the first parallel plants, our knowledge of the gradual or sudden dematerialization of particular plants is still rather sketchy. We know that the two botanies are branches of the same original "tree," but when and how the split took place is for the moment the subject of vague hypotheses based on a few somewhat mysterious discoveries.

On November 28, 1972, exactly a year after the great find at Tihir El, Boris Chersky and Johann von Wandelungen of the University of Freibourg were working not far from the Tiefenau Valley when they unearthed some fossils which might represent the stage of development immediately preceding the mysterious mutation by which the *Tirillus vulgaris* became the first parallel plant. The PL. VII fossils portray a plant rather like an onion, but which is almost certainly a tiril with a large bulbous root. The bigger of the two fossils bears the imprint of a single *Tirillus bulbosus,* as it is now called, while the smaller one, the famous "Hochstadt fragment" quite clearly shows a somewhat elongated bulb from which sprout two common tirils.

The discovery of a single-stemmed plant with a bulbous root and dating from the Erocene era would in any case have been an exceptionally interesting item of scientific news. What made the discovery absolutely sensational was the fact that the tirils have all the features of parallel plants while the bulbs are clearly to be assigned to normal botany.

The studies later carried out by Spinder into the nature of the cellular tissue, the physiognomy of the single cells preserved in a very thin layer of carbon, as well as the analysis of the remains of cellulose filament, leave no doubt as to the normal plantness of the bulbs. But the fragments of tiril are absolutely identical with those of the Ham-el-Dour desert. They give no sign of any organic quality whatever, and in spite of the perfection of the imprint they betray not the slightest alteration which might be attributed to the normal vital functions of an ordinary plant. They are totally lacking not only in organs but in any kind of cytological structure. Their substance, if one can use such a term, must have been an immobile continuum, even in the subatomic state, and insensitive to impulses of any kind. Spinder has no hesitation in regarding these two fossils as paleontological evidence of the moment when parallel and normal botany went their different ways.

However, there are many unanswered questions, and exactly what internal mutation or external conditioning could possibly have caused such a strange evolutionary anomaly is destined to remain a disquieting enigma for some time to come. If we are now in a position to analyze matter and measure time even in the earliest

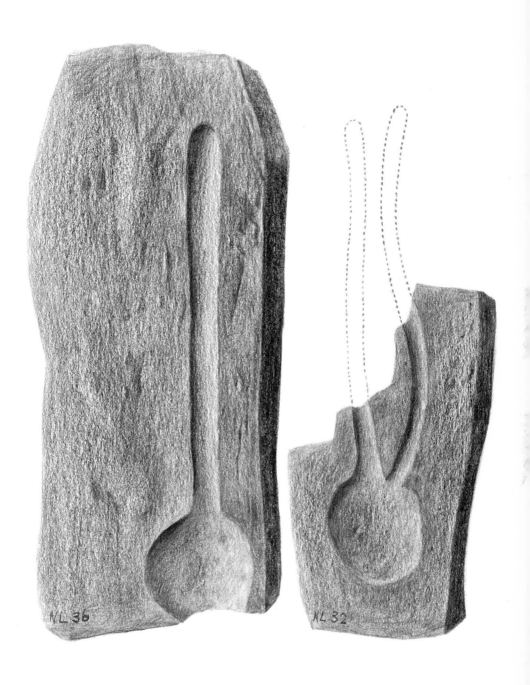

PL. VII Fossils of the bulbous tiril

dawn of the history of our planet, we unfortunately do not possess the means of analyzing non-matter and measuring non-time. One of the great unknowns of parallel paleobotany involves the dating of specimens. The fact that, in a sense, the plants are themselves fossils would surely seem to make it easier to assign them to particular geological periods. But unfortunately this is not the case. Normal fossils are dated by examining objects found in the same environment and by comparing them with the fossils of neighboring organisms. But parallel plants represent a case of substitution, a metamorphosis that at the moment of its coming into being obliterates the previous existence without trace. Scientists generally support the hypothesis that the first parallel plants appeared, give or take a few millions of years, at the beginning of the second half of the post-Plantain era. But we know that plants which came within the normal botanical repertory of our forefathers have undergone parallelizing mutations, and there are even some who speak of processes of dissubstantialization going on at this very moment. As we see, the period of time in which the phenomenon takes place is a very lengthy one and for the time being does not allow us to generalize.

Theoretically, the only certain method of dating a parallel plant would be a radioactive analysis under conditions of thermic saturation, but so far the matterlessness of the plants has proved an insurmountable obstacle. The results obtained by Boris Kalinowski from his carbon 16 test seem to raise hope that in the not too distant future we will have a reliable system for dating all species of parallel plant. The success of the method would also be an important step forward in the study of normal botany. After all, a parallel plant is nothing but the reconcretion of a normal plant at the instant of the sudden and final stoppage of its ontogenesis.

The fossil tirils and those from Tiefenau are the only true fossils which parallel botany has to its credit. They do not seem much, but when one thinks of the unsurmountable obstacle which matterlessness must present to the normal processes of fossilization, their discovery seems little less than a miracle.

Concretions of parallel plants have been found in various parts of the world, but though these objects are doubtless of great importance they are not to be considered real fossils.

In the southern Urals a team of Russian speleologists recently discovered an important stratum in a cave at a depth of 820 meters.

It is said to be rich in fossils of the Erocene era. According to a statement issued by the Paleontological Laboratory at Briskonov, where the specimens are being studied, these include two fossils of the woodland tweezers in a perfect state of preservation. Both Morgentsen and Spinder are of the opinion that these are simply bifurcated forms of *Apsiturum bracconensis,* but they have postponed any definitive judgment until the Soviet government has given them permission to examine the specimens themselves.

MORPHOLOGY

The difficulties of applying traditional methods of research to the study of parallel botany stem chiefly from the matterlessness of the plants. Deprived as they are of any real organs or tissues, their character would be completely indefinable if it were not for the fact that parallel botany is nonetheless botany, and as such it reflects, even if somewhat distantly, many of the most evident features of normal plants. These features or qualities must be seen in the light of the concept of botanicity ("plantness"). For parallel plants, which often possess no other reality than mere appearance, plantness is one thing that enables us to recognize and describe them, and, to some extent, to study their behavior.

What, then, do we mean by plantness?

In substance it is the ideative gestalt, the aggregate of those morphological characteristics which make plants instantly recognizable and placeable within one single kingdom. In other words, it consists of those recognitive elements that make us say of a thing, "It is a plant," or "It looks like a plant," or even "Look, what a strange plant!" This last exclamation, incidentally, gives some idea of how strongly identifiable are the formal characteristics that distinguish plants from all other things on earth. But the process, which seems so elementary, is in reality rather a complicated one. It involves not only the morphological characteristics of the plants and our own possibilities of perception, but also the whole of our complex and ambiguous relationship with nature. Plantness is in fact no more than a particular aspect of the larger concept of organicity, a basic quality common to everything in nature, and the one that usually sets an immediate and unmistakable stamp on outward appearance.

C. H. Waddington, former director of the Institute of Animal Genetics at Edinburgh University, is one of the few scientists who have attempted to describe the formal difference between the products of man and those of nature, relying for evidence not only on the *Aulonia hexagona*, a single-celled organism, but also on the sculptures of Henry Moore and Barbara Hepworth. In an essay on the nature of biological form he puts the problem this way:

> If one found oneself walking along the strand of some un-
> known sea, littered with the debris of broken shells, isolated
> bones, and old lumps of coral of some unfamiliar fauna,
> mingled with the jetsam from the wrecks of strange vessels,
> one feels that one would hardly make any mistakes in dis-
> tinguishing the natural from the man-made objects. Unless the
> churning of the waves had too much corroded them, the odd
> screws, valves, radio terminals, and miscellaneous fitments
> even if fabricated out of bone or some other calcareous shell-
> like material, would bear the unmistakable impress of a human
> artificer and fail to make good a claim to a natural origin.
> What is this character, which the naturally organic possesses
> and the artificial usually lacks? It has something, certainly, to
> do with growth. Organic forms develop. The flow of time is an
> essential component of their full nature."[1]

At first sight the growth factor mentioned by Waddington would seem to be a valid touchstone, but in actual fact it does not really explain our instant ability to tell natural things from human ones. Growth is a vital process, of course, but it takes place over long periods of time, and the morphological changes involved are at the subcellular level, invisible to the naked eye. We do not *see* growth, we simply *know* from previous experience stored in our memories that something *has grown*.

The Hungarian biophilosopher Kormosh Maremsh, in his critique of Waddington's theory, observes that if growth is in fact a touchstone for differentiating between the things of nature and those of man, we are going to find it hard to explain *decrease*. In a particularly brilliant passage, he compares a pebble to a billiard ball and underlines the paradox that while both have reached their final form by a gradual reduction of volume and the simplification of their original forms, the pebble (made of inert material) is still recognizable as a thing of nature while the ball (made of ivory, a living substance) is quite clearly an artifact.

What then is the perceptive process by which, without a moment's hesitation, we tell natural things from the things made by man? What exactly is this quality of organicity that we attribute to the first and deny to the second?

In 1778 Ebenfass (*The Living Machine*) was the first to introduce the word *organisch* when referring to living organisms. For the German philosopher the term had an absolutely precise function: to describe a complex of organs arranged harmoniously. But little by little, by analogy or semantic shift, the word took on other and always broader meanings which became increasingly difficult to define. Nowadays we do not think twice about using it to describe the style of a house, the quality of a line, the shape of a swimming pool. But in general we might say that organicity is the quality which typifies the forms of nature and which is lacking in the products of man.

The problem of comparing nature with artifact was already recognized and discussed, though rather superficially and always within the sphere of aesthetics, by a number of Greek philosophers. But it was only many centuries later, with the Enlightenment, that the emergence of a rudimentary scientific technology enabled it to become the object of a more thorough analysis. That it was a topic of the moment in the early nineteenth century is clearly shown, if only by implication, by an Eskimo legend retold by the Canadian ethnologist Philip Welles (*Men and Myths of the Northwest*, Vancouver, 1842).

Welles, who lived for many years with the Inklit and Tawaida Eskimos, describes the legend as "a modern fairy tale inspired by contact with the Canadian merchants offering manufactured goods such as balls, glasses, beads, mechanical toys and even watches in exchange for skins, ivory and whale oil." The legend was told to him by the shaman of the village of Foipù, at the foot of the Kwapuna mountains. Here it is:

When the god Kanaak wished to create life on earth the first things he invented were sickness and death, then the ferns, the holm oak and the other trees. Then he invented the bear, the whale, the snow cricket, the beaver and the other animals. Finally he invented man, and he taught him to make things,

and to make them in his own image, imperfect. And man
made things in this way, and they served him most perfectly.
He made the kayak like the pod of the Took tree, and with bones
and the fibers of plants he made fish hooks, harpoons and
nets. He dressed in the skins of the white wolf and from the
claws and teeth of the bear he made necklaces and belts. But
one day man discovered that by rubbing one stone against another he could imitate the song of the snow cricket; and he did
so. But one of the stones was harder than the other, and after
he had been rubbing for a while man realized that he had
made a perfect sphere. When he saw it, man realized that he
had sinned against the god Kanaak; and he was afraid. He got
up guiltily and tried to hide the sphere in the hollow trunk of
the tree which he was leaning against, but it slipped out of his
hand and started to roll away. The man ran after it, faster and
faster. Kanaak saw it, but did not stop it. As a punishment he
made man run after it until he disappeared into the endless
darkness of the Kwapuna mountains.

"And he is still running after the perfect sphere," was the ironic
comment of Welles, anticipating by a century our own objections to
an industrial-consumer society.

The first to contrast the notions of nature and artifice, not simply
from the conceptual, intellective, and moral standpoints, but chiefly
from the phenomenological point of view, was Kormosh Maremsh.
In his study of organicity, *Perception and Nature,* a work of fundamental importance to both the study of biology and the understanding of art, he arrives at the following definition of organicity by
means of a long and meticulous analysis of technology in which he
traces its whole evolution: "the continual struggle of man to
dominate the chaotic fatality of nature, to make it comprehensible
and foreseeable." Starting from the day on which a man for the first
time picked up a stone to keep and use ("the first real human
gesture"), he describes the course and the gradual transformation of
primitive tools and household objects into the industrial and con-
sumeristic equipment of our own days. He sees in the development
of manufactured objects the slow penetration of a language which
little by little alters their function, producing more and more abstract forms. While the things of nature have no function other
than to exist in themselves, one which they express morphologically

by their unity of appearance (Portmann's "self-presentation"[2]), manufactured things need two efficiency factors, one mechanical, the other symbolic. "On a par with mechanical functionality," writes Maremsh, "man always tends to choose for the things he makes the solution that is richest in message, the most loaded with meaning. And thus the language of objects has undergone a development comparable with that of the language of words: it already has its own grammar, syntax and rhetoric." And again: "The history of technology shows us the gradual transformation of things of use into objects of possession, of utensils that are eloquently mechanical into ritual and abstractly linguistic instruments."

While Maremsh sees this evolution as the result of economic and political struggles, the psychologist Wolfgang Keller thinks he can perceive in it some of the psychological causes inherent in the ideative process. He speaks in particular of what he calls "the geometric impulse," which is in fact the title of a recently published book of his.[3] Drawing the distinction between instinct and impulse, the German psychologist writes: "While certain animals possess a rudimentary geometric instinct, usually concerned with the standardized production of a single object (spider's web, honeycomb), only man, gifted with imagination, possesses the capacity to project, to verify, and the irresistible impulse to realize things in concrete terms."

He goes on to explain that "the vision of the imagined thing is as a rule primarily an interpolation, stylized and gestaltic. Its forms appear in the mind not by the gradual and systematic addition of one part at a time, but by the simultaneous emergence of a whole. This ideative process is characterized by an alternation of propositions which is bound to culminate in the choice of the form which, in opposition to the chaos of the real world, best represents a clearly discernible order such as that of geometry.

"The 'design,' which is a proposal to render imagined objects concrete, tends to choose out of all imagined forms the one that is most easily perceptible as gestalt, as an organized geometric whole. This impulse toward geometry, already institutionalized in the profession of 'designer,' is responsible for the proliferation of ever more abstract objects, increasingly in contrast with natural forms."

Keller then observes that the geometric impulse is not confined to the creation of objects but also seems to dominate our interpretation of everything round us, including nature. Unable to accept the chaos which is characteristic of free forms of nature, man imprisons them in definable and measurable schemes, his own body being no exception to the rule.

The result of a lifetime devoted to the measurement of nature, D'Arcy Thompson's vast and comprehensive 1100-page volume *On Growth and Form*[4] gives us all possible and imaginable aspects of mathematics and geometry as applied to living forms, from the growth of Belgian children to that of herrings, from the curves of horns, teeth, and claws to the parabola described by a hopping flea, from the shape of a waterdrop to the arrangement of leaves on a stem. Designs, diagrams, outlines, and simplifications transform living things into models of the most rigorous symmetry.

In his excellent little volume *Natura e geometria*, Aldo Montù confesses: "The observation of facts leads to an instinctive rebellion against a geometric simplification and unification that does not make allowance for single events—but in reality there is order in the whole and a great liberty of variation in the particulars, and this determines the harmony of all relations."[5] But then, not making allowance for single events, Montù goes on to circumscribe and imprison the free forms of shells, flowers, and leaves in squares, circles, rectangles, triangles, ellipses, and hexagons. Beneath the geometrical figures, however, the photos reveal the chaotic outlines, the chance distribution of spots, the rebellious excrescences, the veins of irregular size and spacing, all of which not only characterize individuality but comprise its *sine qua non*, that impetuous disorder which eludes measurable generalizations, as do the things of nature.

It is obvious that when we are dealing with the appearances of things and our perception of them, diagrams are just as useless as words. After even the most effective geometric analysis or verbal description the images we seek to evoke remain nebulous and unstable, likely to be deformed by the least touch of interpretation.

Aware of these difficulties, Maremsh has supported his observations with figurative examples of theoretical but real situations from which, by means of the direct comparison of natural and manufactured objects, the meaning of organicity emerges with the greatest clarity.

While recognizing that "it is not possible to teach anyone to read organicity, but luckily we read it as naturally as we walk," the Hungarian philosopher involves us directly in the reading of partic-

(a)

(b)

(c)

Fig. 7 From *Perception and Nature* by Kormosh Maremsh

Fig. 8 From *Perception and Nature* by Kormosh Maremsh

ular cases in which different levels and degrees of organicity are put face to face. From an examination of the examples, some of which are here reproduced, the concept of organicity gains solidity, free from the restrictive exigencies and misunderstandings of verbal definitions.

Taking Waddington's theory as his point of departure, Maremsh imagines himself on a beach, looking at pebbles. Although the action of the water has blunted the points and worn down any sharp edges, the shapes remain clearly organic and not susceptible to any easy geometrical definition (Fig. 7a). Even in a group of exceptionally regular pebbles, a perfectly spherical object immediately leaps to view as a man-made thing. Any child would recognize a billiard ball as a billiard ball, even if it had been worked on by the action of the sun and the waves, the salt and the grinding sand (Fig. 7b). In the same way, we will have no difficulty in recognizing a pebble in the midst of a group of billiard balls (Fig. 7c). But Maremsh points out that if one of the balls were split in two we would know it as a billiard ball only "by association." Among the pebbles this split ball would be hard to distinguish as a man-made object because of the aggressive "organicity" of the fracture. "Continual wear," observes Maremsh, "gives human products a certain degree of organicity."

Three versions of a branch with a fruit hanging on it (Fig. 8) form what is perhaps the Hungarian philosopher's most noted demonstration. The illustrations clearly show how eloquently both organicity and inorganicity survive the most anomalous contexts. In the first version the situation is completely natural. Although it is not possible to make out the species of plant, and although it is only a fragment of the whole plant, the branch and the twig and the fruit do nevertheless compose a whole which indubitably possesses organicity (plantness). The branch in the second illustration, however, is immediately read as a stick to which a real twig and real fruit have been inexplicably attached. The third illustration is a man-made object which we interpret as a stylized representation of a branch bearing a fruit.

Here (Fig. 9) are some of the famous leaves with which, as in the example above, Maremsh not only shows the characteristics that mark off the organicity of things of nature from the inorganicity of human products but also clearly demonstrates some of the most typical features of plantness. Maremsh here illustrates several of the salient points in the theory he developed in his study *The*

Fig. 9 Maremsh's leaves

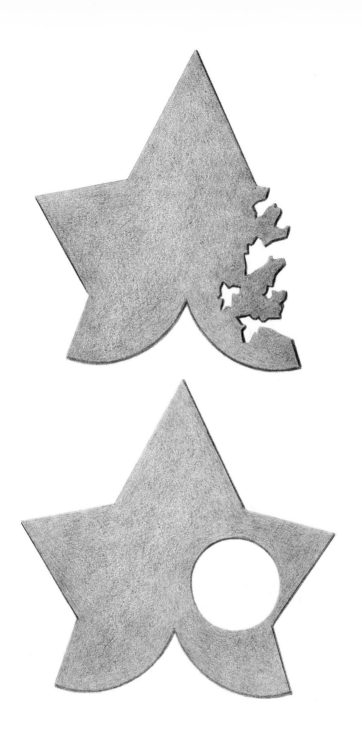

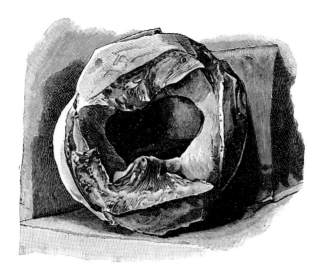

Fig. 10 Maremsh's bagel

Pathology of the Object, especially regarding the destructive action of man and nature on natural and man-made things respectively. The leaves in these examples are immediately recognizable either as organic or as artifacts (or as we usually say, "true" or "false"). Of this series of illustrations the most interesting are those which show the results of human action on a "real" leaf and that of nature on a man-made leaf: both are situations which, in spite of their patent absurdity, reveal how easy it is to distinguish between organic and inorganic forms.

Kormosh Maremsh used the example of the bagel to show that in spite of considerable alterations in the direction of organicity produced by the action of yeast and fire, the manufactured object loses little of its evident human origin (Fig. 10). There are obviously cases in which the effect of natural forces is so violent that it obliterates the original forms of man-made objects, while in the same way human manipulation can end by completely destroying organic forms (as, for example, in the transformation of raw materials).

Turning to aesthetic problems, in which the word "organic" has taken on particular significance, Maremsh gives us the example of four lines (Fig. 11), of which the first was drawn mechanically by

man and is clearly inorganic. The second line is disturbed by a fact
of organic origin (tremor, error, failure of the machine). The third
line is a characteristic detail from a drawing by the American artist
Ben Shahn. Many art critics use the term *organic* to indicate the
artist's intention to approach an autonomous organicity in his drawing by means of deliberate hesitations, errors, and imperfections. It
is an exceptional and very complex situation in which the artist
expresses our ambiguous relation to nature, and in fact sets out to
"replace" nature. In the fourth figure we are shown the lines formed
by the cracks in an asphalt pavement. According to Maremsh these
lines represent "the reacquisition by the soil beneath the pavement
of the organicity which man has attempted to suppress."

The examples given by Maremsh refer as much to the formal as
to the textural qualities, but in spite of the ingenious efficacy of his

Fig. 11 Maremsh's lines

method they can provide only partial answers to the basic questions. When we look at the illustrations given by Maremsh we often have reactions or make choices that are not explicable except in terms of the knowledge and experience that have accumulated in our memories. The associations, direct or taught, which we had with the world of nature (or of men) during our early childhood have left us with so much intellective and perceptive information that they alone would enable us to find our way in the intricate landscape in which we live.

But this inadequately explains our ability to distinguish not only natural things from manufactured ones, but also pebbles from shells, birds from fish, men from monkeys, and plants from all the other things on earth. Although our world is infinitely more complex than the world of animals, we cannot attribute our gift for generalization simply to our human characteristics. "It is not difficult," concludes Maremsh, "to perceive the difference between the organicity of natural things and of man-made objects. But we must at the same time admit that neither my dog Fidel nor my goat Caroline has ever made a mistake."

Taking the work of Maremsh as his starting point, the morphologist Adolf Boehmen has made organicity in botany his chief concern. "Plantness," he explains, "is nothing but the generalization of those particular organic qualities which plants have in common." He goes on to list these aspects as immobility, verticality, color, and texture, and he examines their nature and meaning in great detail in his book *Notes Toward a Vegetable Semantics*. Boehmen goes particularly deeply into the examination of textures, which in some cases provide a determining key to perception. Especially well-known is his experiment in the interpretation of pictures of a lemon, described in the book and illustrated with the original photos used in the test. The experiment consisted of naming a certain fruit represented in several very clear photographs. One of these was an accurate color reproduction of a lemon, seen from the side. Another showed the same fruit in black and white. In the third the lemon was colored orange, while the fourth showed an orange colored lemon yellow. Obviously the reading of the first photo presented no difficulties, while the picture in black and white produced similar results (97 percent of those interviewed recognized it as a lemon). But the orange-colored lemon was interpreted by 86 percent as an orange, and the lemon-colored orange by 91 percent as a lemon.

From this particular case Boehmen deduced that in our assessment of plantness we give priority to color (which can easily deceive us) and only in its absence do we turn our attention to form and texture, which either by analogy or by previous experience will reveal the object specifically as a lemon, less specifically as a fruit, and generically as belonging to the vegetable kingdom.

The experiments of the Austrian morphologist are of particular importance to the understanding of parallel botany, which in the vast majority of cases presents only form and texture as identifying characteristics. Its recent discovery precludes the possibility that our immediate recognition of it as part of the vegetable kingdom was in any way conditioned by previous direct experience. The absence of color, the frequently disquieting contexts, and the morphological oddities of the separate parts might well obscure our reading of it. In spite of this, the plantness of the whole and of its textural qualities is so evident as to leave no doubt that the parallel plants belong with the other flora of the earth; and only a more detailed study of them will reveal them as parallel.

Among the specific cases examined by Boehmen is one of particular interest, possessing the morphological characteristics of both parallel flora and human artifacts. This is the *Solea* preserved in the little museum at Backstone, Massachusetts. The plant is a reconstruction dating from the end of the eighteenth century and attributed to a certain Franco Casoni, an Italian immigrant of Ligurian origin, who established himself and his family at Backstone in somewhat obscure circumstances. There he worked as a wood carver, making a good reputation in his profession, especially for the decorative carvings of flower motifs which still adorn the interiors of the white aristocratic houses of the little New England town. His model of the *Solea* was made under the direction of an Irish seaman, Dominic McPerry, who declared that he had seen the original plant on an island in the Carades Archipelago. This *Solea*, of which the Laboratorio delle Campora possesses a plaster cast, is one of the most perfect known to us. Its plantness is exemplary, in the sense that it expresses all the most characteristic features of plantness to perfection: the utter verticality, the immobility so extreme that it seems to place it outside time, the organic quality of the protuberances and excrescences, even the marks of disease and the wounds (an interesting case of paramimesis) all contribute, in spite of a strange inadequacy of form, to stamp it unequivocally as a parallel plant, typically lacking in any kind of function or meaning. Although in fact it is made by a man, its presence—so clearly an end in itself—has that mysterious quality of self-presentation that Boehmen terms "*Selbstsein.*"

The Backstone *Solea* is mounted on a base of darker wood, also in all probability the work of Casoni. It speaks as eloquently of its man-madeness as the sculpture it supports speaks of its plantness. It is what an accessory should be, in the sense that it serves to complement another object in a clear functional relationship. This base is round and stands on three spherical knobs, to avoid possible damage from insects or damp, while at the same time guaranteeing the maximum of stability. Three concentric circles, equally spaced, with no function other than the purely aesthetic, give the whole thing a modest pretension that in the context of a small-town museum qualify it at once as a "museum piece." If the *Solea* represents nothing other than itself, the forms of the base clearly portray the usual functions of the human artifact. Plant and base together are an eloquent symbol of the conflict between the two kinds of thing which populate our world.

An organicity of the botanical type is the most obvious and the most general aspect of parallel flora. The absence of organs, functions, matter, and growth prevents us from describing parallel plants analytically. While treatises on normal botany have long chapters on evolution, cytology, nutrition, reproduction, and the growth of plants, parallel botany, matterless by nature, gives us nothing we can analyze except its morphology. But as we have observed elsewhere, the known species are few, and the specimens rare and difficult of access. In the same way a systematic morphology, based on a sufficient number of separate observations to arrive at statistically valid findings, is not possible. Unfortunately we have to be content with the reports and the observations recorded in scientific journals, and even these are sporadic and not always reliable.

Regarding the size of the plants there is not much to add to what is known of botany in general. As for normal plants, there is great variety even within a single species. Naturally enough there are no variations due to growth: growth does not occur in parallel botany, its plants being the result of a permanent stoppage in time. The size of known and documented plants varies from that of the *Ninnola preciosa,* which never exceeds three millimeters in height, to the *Fontanasa Stalinska,* which Muyansky describes as "taller than the

PL. VIII Kumode plants

famous oak in Pushkin Park." There are *Giraluna* ten centimeters
high, while the tallest *Giraluna gigas* in the Lady Isobel Middleton
group measures nearly four meters in height. While the *Solea* does
not in general exceed a meter and a half, we do know of a *Solea
argentea* (the one from Amendapur) which reaches three meters.
Within the same species the greatest variation in size is met with in
the parallel pseudofungus *Protorbis,* which ranges from the few
centimeters of the Indian *P. minor* to the twenty-two meters and
more of the *Protorbis* which compete in bulk with the mesas of
Colorado and New Mexico.

One unusual and rather disturbing case is that of the dimensions
of the *Anaclea* discovered by Kamikochi Kiyomasa of Osaka University. About fifteen kilometers from Nara, the ancient capital of
Japan, famous for its temples and monuments, among which is the
gigantic statue of the Buddha called Daibutsu, there is a picturesque
valley from the floor of which, like a large island, rise seven hills
which are vaguely reminiscent of the arrangement and proportions
of the seven hills of Rome. The collective name for these hills is
Kumosan, from the plant called kumode, similar to the myrtle, PL. VIII
which covers almost their entire surface. In the late spring the
kumode puts forth a violet flower with seven petals, the wonderfully
sweet smell of which attracts millions of bees from every corner of
Yamashima province. The famous honey called gokumodemono
gets its special flavor from these flowers. In the procession with
which the picturesque celebrations of Ura Matsuri begin, the
gokumodemono is borne aloft in a bronze vessel dating from the
eighth century and then poured out onto the feet of the Daibutsu, to
the sound of sacred hymns and prayers.

Kamikochi, one of the most renowned Japanese biologists, was
born at Nara. A devout Buddhist, he goes back each year to Nara for
the Ura Matsuri festivities, and he often retires to a rush cabin in
the valley of Higashi-tani, near the hills of Kumosan, for a week of
spiritual exercises. It was during his retreat in 1970 that Kamikochi
made his spectacular discovery. While he was out for a walk his eye
happened to fall on a cluster of unusual flowers on a hilltop, nestled
among the kumode. They were about a hundred meters from where
he was standing. He was unable to make out their color because
they appeared as black silhouettes against the bright sky, but their
shape seemed very strange. He found it hard to estimate their size
because, apart from the surrounding kumode, he had nothing to
compare them with.

Kamikochi decided to take a closer look at the flowers and started walking toward the hilltop. On the way he realized that something very bizarre was taking place. Unlike what usually happens when we approach an object we have seen from a distance—which gradually appears larger until, when we are near enough to touch it, it assumes its proper dimensions—these plants did not seem to get bigger as the biologist approached them. When Kamikochi reached the hilltop they turned out to be just as small as they had appeared from a hundred meters away.

At first he was inclined to attribute this phenomenon to the long hours of meditation which he had practiced before his walk. But when he repeated the experiment the result was identical. He did it a third time, taking care not to lose sight of the plants for an instant, and after that he was quite certain that as he drew nearer the plants their apparent size did not alter in the least.

A few weeks later Kamikochi returned with a number of his pupils to study the problem, which he called "metrostasis" and which he described in a paper at the botany congress held in Tsuchimachi in 1974. He said that the flowers were of the species PL. IX *Anaclea taludensis*, and measured at the most fifteen centimeters in height. They are completely black, and there is not the remotest doubt that they belong to parallel botany. It is impossible to pick them, as they vaporize instantly on contact with a hand or any other object that is not part of their normal ecological environment.

Kamikochi, though admitting that he was unable to give a scientifically satisfactory explanation of the phenomenon, attributes it to the immobility in time characteristic of parallel plants, and quotes Leibschmidt's law to the effect that "for every immobility in time there is a corresponding immobility in space."

"The type of perspective," he explains, "that reduces the image of a distant object in proportion to its distance from the point of observation presupposes a normal time-space relationship. A change in the fundamental qualities of one of the two elements must of necessity imply a change in the other." If at first sight Kamikochi's argument appears irreproachable, we are led to ask ourselves why it is that the other parallel plants are not subject to the same phenomenon.

A team of neurologists, psychologists, and opticians at the University of Osaka is now working on the problem of metrostasis. It is by no means impossible that certain plants might have disturbing

PL. IX *Anaclea taludensis*

effects on human eyesight. Harold MacLohen, in an article in the Chicago *Times,* reminds us of how recently in human history we have come to accept mere images as reality. "For millions of spectators," he observes, "the leading personalities of our time—athletes, statesmen, pop singers, and scientists—are at most ten inches tall. We accept their rather dubious dimensions without ever being able to verify them in person."

The colors of plants and their morphological characteristics are part of the language in which they carry on their dialogue with the world. It is by these means that they transmit important messages regarding personal identity and survival. The color green, characteristic of the stems and leaves, is a secondary effect of chlorophyll. It expresses the harmonious functioning of the vital processes for which chlorophyll, as an intermediary of nutrition, is largely responsible. When these processes are damaged by pathological conditions or suspended by the seasonal drying-up of the plant, the color alters and signals what is happening.

The function of the other colors, particularly that of the flowers, is more mysterious. While the green informs us of the health of the individual plant, and is therefore a simple affirmation, the other colors are invocations, invitations, questions. They have to do not so much with the survival of the individual as with that of the whole species. As Hamilton puts it: "For plants, by a cruel fate deprived of motion, colors are a silent language of love, desperate and passionate, a language which birds and insects, their winged messengers, carry to distant lovers also ineluctably fastened to the earth."[6]

This English biologist is of the opinion that for parallel plants, "fixed not in earth but in an inert time," the problem of survival does not exist. As a result, color as an instrument or a signal would only be justified as a paramimetic phenomenon, that is, as a trick to disguise their true nature. "When this happens," he adds, "we can assume the existence of an exceptional anomaly, because parallel plants are without any life as it is lived in the flow of time, and they therefore have no need for color." Hamilton's remarks, which at first seem logical enough, contain two basic flaws. To begin with, when he asserts that parallel plants have no color *because they do not need it* he clearly ignores the recent departure from traditional evolutionary theories. Portmann has opened our eyes to the fact that many natural phenomena, traditionally thought to have some functional significance with regard to survival, are, in fact, entirely gratuitous and inexplicable in rational terms. Second, if it is true that we cannot speak of a real color in the case of parallel plants, partly because their surface is just the external limit of an interior, their visibility must nevertheless be expressible in chromatic terms. If the variations in their degree of opacity and the indefinable nuances of black sometimes seem like a lack of color, a void in the colored world that surrounds them, in reality these characteristics are positive and typical of parallel plants, directly connected with their mode of being. It is not easy to describe and explain these characteristics, for they are as elusive and ambiguous as the plants themselves. Jean Parottier writes: "While the colors of normal plants share the hard certainty of sunlight, those of the parallel plants seem to hang in the dreamlike ambiguity of the darkness of night." And again: "The colors of these plants aspire to the condition of night. And as it is hard to find a pure black even on the darkest night, so it is with the parallel plants."

The gamut of blacks in parallel plants ranges from *"tête de nègre,"* as warm and mysterious as a bronze by Rodin, to the cold and hostile black which Delacroix called *"bois brûlé."* But it is the strange sheen of these different nuances of black that gives the parallel plants their curiously matterless and sometimes almost spectral appearance. It is like a skin of light within the pigments, causing both the shadows and the strongly lighted areas to lose their outlines. The surface of the parallel plants more than anything else resembles the patina found on ancient bronzes, which is also difficult to describe, not because it has no color but because the slow wearing away of time has mitigated its arrogant aggressiveness, that presumptuous self-confidence typical of man's artifacts when they are new, and of the things of nature when they are young.

The discovery which Theodor Nass made accidentally, and which aroused a great deal of interest some years ago, has revealed some mysterious and disquieting aspects of the chromatic features of parallel plants, aspects that may one day lead us to a fuller understanding of what he calls "parallel black." During a period of research at the Laboratorio delle Campora, the famous Swiss scientist inserted part of the *Solea fortius,* one of the most valuable specimens in the great Chianti collection, into a block of polyephymerol, a new plastic with as yet unexplained properties of refraction, for which it is widely used for the lenses of Bonsen refractometers. If polyephymerol is cut and set at a certain angle it reveals characteristics similar to those of laser beams. Nass in fact was using it to make three-dimensional measurements of the growth-spiral visible all around the *Solea,* which Nass suspected might show analogies with the DNA spiral.

When he had inserted part of the *Solea* into the cube he found to his astonishment that inside the plastic this plant, normally one of the blackest of all parallel plants, appeared vividly colored. Nass, completely unable to give a logical explanation for this, put forward the hypothesis that the dark patina of the plant could in fact be merely the upper layer of a number of superimposed colored layers, a kind of screen which usually conceals the pigments and which with the aid of polyephymerol we are able to penetrate. The *Solea* with its center encased in the plastic cube was shown at the Parallel Botany Exhibition arranged in conjunction with the Offenbach Conference of 1973, where its mysterious chromatic behavior attracted the attention of the world's press. In an interview with the *Frankfurter Tagesblat,* Nass among other things revealed that polyephymerol contains a derivative of amitocaspolytene, a rare and highly poisonous substance. This imprudent statement gave rise to an inquiry on the part of the German authorities, which is still going on. The chairman of S.I.M.A., producers of polyephymerol, gave a press conference in which he assured journalists that all due precautions had been taken to protect the health of workers and laboratory staff, and that the final product was perfectly inert and harmless. By way of demonstration he showed a photo of himself standing beside his three-year-old son Johann, who was holding in his hands a shapeless lump of polyephymerol. At a second press conference Nass declared that further tests carried out by him into the toxic nature of the material had shown completely negative results.

Following the Offenbach Conference, Nass obtained a grant from the Geremia Pirelli Foundation to enable him to continue his experiments, extending them to other parallel plants and even to species of normal botany. The first results, though varying greatly in intensity, were similar to those obtained with *Solea fortius.* With normal plants, on the other hand, the phenomenon did not occur at all. Only one part of the stem of a Princess Grace rose under the plastic showed a slight bluish tinge which, according to Nass, might

be the beginning of a mutation, the prelude to a possible parallelization of all roses. This strange phenomenon, which is now known as
the Nass chromation, has not yet been satisfactorily explained.
Nass carries out his experiments in great secrecy, and is retiring
and evasive even with his colleagues in the laboratory. There has
also been no explanation of how Professor Vanni,[7] director of the
famous Italian laboratory, who has a reputation as a prudent scientist and a meticulous administrator, could have allowed Nass to
experiment on the rarest and most precious of his whole collection
of *Solea,* thereby risking its total disintegration. Maybe these and
other questions will be answered at the forthcoming conference on
parallel philosophy, due to be held in Tokyo in 1978. Both Nass and
Vanni will be among the speakers on that occasion.

PART TWO

THE PLANTS

THE TIRILLUS

The vast majority of parallel plants are found in isolation or in small groups that rarely exceed a dozen in number. The tirils, like the woodland sugartongs, are by nature gregarious and live in dense groups that sometimes cover large areas. Examples of this are found especially in the tundras of Ackerman's Land, not far from the Borloff Straits, where endless fields of tirils stretch as far as the eye can see, disappearing only into the frozen mist of the Arctic horizon.

Of all parallel flora the tiril has the widest and most varied PL. X geographical distribution. From the Arctic tundras to the Peruvian Andes, from the Siberian steppes to the mesas of Patalonia, from the Omar Delta to the shores of the Gulf of Good Friends, scholars and travelers interested in botany have recorded the presence of *Tirillus vulgaris* and of its more aristocratic varieties: *T. major, T. tigrinus* (Fig. 12), and *T. tihirlus extinctus* (fossil variety).

Its modest form, as elemental as that of our grass, has been known since time immemorial, and was recorded as long as three thousand years ago on the petroglyphs of the Moqui Indians who lived on the Fremont River in Utah. Theocrastus mentions it in his "Rural Discourse" and a Mannonite manuscript cites the tiril as an example of perfect and peaceful communal life.

The *Tirillus vulgaris* or common tiril is the prototype of the species, measuring from 18 to 26 centimeters in height and approximately a centimeter thick. It lives in dense groups, sometimes with as many as four thousand individual plants to the square meter. Often the plants are so closely packed together that in preparallel conditions the sick plants found it impossible to wilt, and died in an erect posture, supported by their neighbors.

In parallel botany the tiril is the only species in which normal examples can be found next to petrified or fossilized specimens. At Ampu-Chichi in Peru, about ninety kilometers from the highest point of the Chimu-Pichu, the paleobotanist Edward H. Kinsington has discovered whole fields of fossilized tirils which apparently perished in some epidemic; all are in a perfect state of preservation, and all in a vertical position. "They look like fields of breadsticks," writes the Australian paleontologist, who as a young man lived for a while in Turin, attending Onofrio Pennisi's famous university courses in myopaleontology. He was therefore familiar with the Piedmontese *grissini* of that time, though in fact it was during the First World War, when flour was not of the best quality.

The "black" of the tiril is among the most colorful of all parallel plants. In the few tests that have been made with the Fersen chromometer, readings in F-vibrations from 82° F. to 112° F. have been obtained, which is a fairly considerable range if we come to think that the color of black plants is, basically, one and one only. These readings are not much lower than those obtained from the leaf of *Frenemona taliensa,* the most highly colored of all the parallel plants.

It may well be asked why we are not in possession of more accurate data on the color of a plant with such wide geographical distribution. There are many reasons for this: the bulk and weight of the chromometer, the uncontrollable variations in atmospheric pressure, the fluctuations of ozonoferous density. But the main reason, mentioned earlier, is the impossibility of moving the plants from one place to another. This difficulty exists not only for tirils, but for all parallel plants.

The Laboratorio delle Campora has carried out numerous experiments in the transport of these plants, but unfortunately none has yet met with success. However, Marcello Vanni, director of the laboratory, with the help of Valerio Tarquini, is at present perfecting a new type of pressurized equipment which it is hoped will produce positive results in the foreseeable future.

As regards the obstacles to habitat-exchange, we ought perhaps to repeat a few general observations. As a direct result of their matterlessness, parallel plants submit fairly gracefully to verbal description but are decidedly hostile to any method of documentation that threatens to duplicate what in them is mere appearance, especially through the medium of a false "reality" such as photogra-

PL. X Tirils

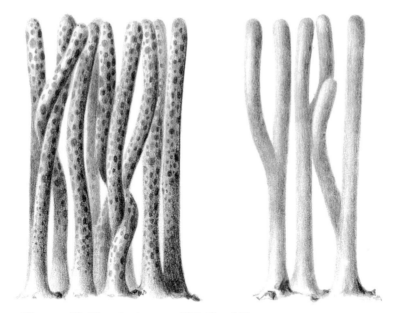

Fig. 12 *Tirillus tigrinus* and *Tirillus bifurcatus*

phy. There have been innumerable unsuccessful attempts to photograph apparently accessible plants, even though the use of the most refined equipment seemed practically to guarantee success. Thus, except for these inclusions and a few presumably direct casts, our chief sources of information are unfortunately verbal reports and three-dimensional reconstructions modeled either from memory or from drawings done from life.

All the same, the tiril is the best documented of all the parallel plants. Since 1972 the Tirillus Society of America has been informed of as many as eleven newly discovered beds, six of them in Central and South America, thanks to the expeditions of Kinsington and of Roger Lamont-Paquit of the Parabiology Institute of Catarás. Of the remainder, three are in Africa and two are in Siberia.

With the exception of *T. bifurcatus* (Fig. 12) of Jakruzia and the anomalous, autoparasitic specimens found by Karovsky on the Arkistan Plateau, all known varieties of tiril possess similar morphological characteristics. If they differ from the prototype *T. vulgaris,* it is more in behavior than in form. There is, it is true, the dwarf variety that is found on the edge of the Great African Rift, near Lake Kivu in Ruanda, but it is possible that we cannot perceive its true dimensions and that its behavior is similar to that of the *Anaclea taludensis* observed and studied by the Japanese biologist Kamikochi.

We here list the fourteen varieties of the prototype *Tirillus vulgaris,* together with the location where they were found (reprinted from the yearbook of the Tirillus Society of America).

T. odoratus	Mexico	Sierra Madre
T. silvador	Peru	Ampu-Chichi
T. oniricus	Ecuador	Cordillera Real
T. mimeticus	Brazil	Carimà
T. tigrinus	Argentina	Quequà
T. parasiticus	Brazil	Rio Samonà
T. tirillus	Siberia	Jakruzia
T. tundrosus	Alaska	Helmutland
T. major	Pacific	Western Patagonia
T. omarensis	Asia	Omar Delta
T. tihirlus extinctus	Iran	Sarab Bainah
T. minimus	Ruanda	Lake Kivu
T. bifurcatus	Siberia	Jakruzia
T. bulbosus	Germany	Bavaria

Tirillus oniricus

T. oniricus was first discovered in Ecuador by the American poet John Kerry ("Solitary John"),[1] who passed on the information to his biologist friend Roger Lamont-Paquit[2] when he learned about the research which the latter was carrying out in the field of parallel botany. It is to this French scientist that we owe all our documentation of this strange plant, the fruit of long and patient study during his expeditions along the Cordillera Real. On the Cucutà Plateau, at the foot of the great volcano Chimborazo, he found the little colonies of tirils described by Kerry, hidden amidst the endless stretches of *Ostunia fluensis* (the *boho* of the Javaros Indians). The *boho* produces the black and perfectly triangular seeds which Indian children use for playing *salari,* a game which in recent years has become quite popular in Scandinavia in a colored plastic version

Fig. 13 The *Ostunia fluensis* and its seeds (above) and the game of Skååp-Skååp

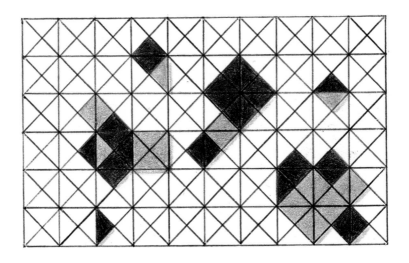

called Skååp-Skååp (Fig. 13). Lamont-Paquit tried to photograph
the tirils of Ecuador with a special Roemsen lens from a distance of
one kilometer, but unfortunately where the tirils should be the
photo shows only a white blur.

T. *oniricus* seems to all appearances a normal tiril, no taller or
more robust than its more common and accessible brethren. The
disquieting feature of this plant is that its image apparently lodges
in some yet unknown corner of the memory, where at irregular
intervals it makes its presence felt. After Lamont-Paquit first saw a
field of these tirils he suffered from a strange malaise that he at first
put down to the effect of the altitude. He sat down on a boulder,
expecting to recover, but the image of the plants continued to recur,
to flood into his mind in an almost stroboscopic sequence, but at
unpredictable intervals. When he succeeded in directing his
thoughts to other memories, even in the distant past, everything
became normal again and the strange feeling suddenly vanished.
Gradually he learned to control the phenomenon, forcing himself to
think first of the tirils and then of something else, thus controlling
also the sensation of slight dizziness that always accompanied the
appearance of the mental image of the plants.

The neurologist Theodor Kinderstein, noted for his experiments
with hallucinogenic substances, worked for some time with Lamont-
Paquit. In his opinion, the phenomenon is similar to that produced
by the bombardment of the retina with the rays of a flame obtained
by burning Toxcaline. It seems that *T. oniricus* has this characteristic, fortunately rare in nature, caused by occasional fractures in the
light waves that emanate indirectly from its surface. The resulting
intermittent appearance of the image in the memory has not so far
been satisfactorily explained. Lamont-Paquit, who has undergone
numerous neurocephalic tests at the Institute of Neurology at
Lyons, has learned not to remember the *Tirillus oniricus* which he
discovered, for strictly preestablished periods of time. In this he has
been greatly aided by the Indian guru Ajit Barahanji, who has given
him exercises in mental deconcentration and selective dethinking.

Tirillus mimeticus

As parallel plants—unlike their sisters on this side of the hedge (we
resort to Dulieu's memorable phrase)—are not cursed with conflicts

and struggles for survival, it is difficult to understand how it came about that on Carimà island in the estuary of the Rio de las Almas there should be a tiril so perfectly camouflaged that to all intents and purposes it is invisible. Lamont-Paquit, who discovered *T. oniricus* in Ecuador, describes *T. mimeticus* as slightly shorter than the average for normal tirils (about twelve centimeters), but having the same diameter. It is regular in form, without scent, and has a

Fig. 14 *Tirillus mimeticus* and stones

smooth mat surface. It lives in small family groups of not more than sixteen individuals, maintaining a constant distance of three centimeters from its nearest relatives. Lamont-Paquit gives us this information as pure deduction, because the plant is so perfectly camouflaged in its habitat of black volcanic pebbles that it is not discernible by normal visual means.

Lamont-Paquit was told about this tiril by the Indians of Quahac, not far from Carimà and the only island of the archipelago still inhabited. Late one summer afternoon he was taken to Carimà by the shaman and a number of young Indians, and there he was shown a group of tirils which in the red light of the sunset ought to have revealed some vestiges of form. But the biologist saw nothing. With the help of the Indians he thought he succeeded in touching a few plants, which surprised him very much, in view of the well-known vulnerability of all parallel plants to the human touch. He took a vast number of photographs using different lenses and types of film, and unlike that of his photographic misadventures in Ecuador the result was, in a sense, positive. In the enlargements one can clearly see every detail of the rounded stones and of what must be the T. *mimeticus,* though so far it has not proved possible to tell the one from the other (Fig. 14).

In reply to Lamont-Paquit's questions on the matter, the Indians provided a mythological explanation of the phenomenon. When a shaman dies, they said, his soul dives into the waters of the estuary and swims under water to the islet of Carimà. There it emerges and stands invisible among the stones, watching the actions of its living brethren. When an Okonò is tempted to commit a sin of violence, the soul-tiril (Tu'i-sa) shoots an arrow that wounds the guilty one, punishing him with fever, pains, and vomiting. In point of fact the Okonò never confess their bodily pains, for fear that they might be revealing secret desires for violence. When they are ill these peace-loving Indians hide away and suffer in silence, and when the breeze is blowing in the direction of Carimà they propitiate the vengeful
PL. XI souls by making little boats out of dried Saklam leaves, filling them with colored feathers and Cufolà seeds, and sailing them off toward the isle of the tirils.

During his last visit to the island of Carimà, Lamont-Paquit saw a small flotilla of these tiny boats arrive on the beach. Fearing that a wave of violence or an epidemic had struck the neighboring island, the scientist packed his bags and left in a hurry.

PL. XI Propitiatory leaf-boat of the Okonò Indians

Tirillus parasiticus

The parasitic tiril grows on dead tree trunks and branches in certain PL. XII
tropical forests. Of normal shape and size, it differs from other
kinds of tiril mainly in the composition of the groups. The plants
are in fact arranged in line a few centimeters one from another, and
there is always an odd number in each group. The groups are
therefore symmetrical, and this feature is often accentuated by the
central tiril being taller than the rest.

These tirils are not true parasites. The term *parasite* is derived
from the Greek *parà* (beside) and *sitos* (food), and implies the
exploitation of the conditions or qualities of others for one's own
survival. This clearly does not apply to plants whose existence is
independent of relations with other organisms or with the environment, and whose life cycle, if we can use such a term, occurs in a
motionless time which ensures the permanence of their bodily
forms.

The name "parasitic tiril" was used to describe these plants by
Jacob Scheinbach of Hanover University, who certainly had no idea
of the misunderstanding this unintentional baptism would create.
Scheinbach discovered the plants during an expedition through the
Brazilian forests in the region of the Rio Samonà, and in a letter to
his colleague Metzen he used the word *parasite* to simplify what
would otherwise have been a rather lengthy description: "Yesterday
I saw a plant rather like an asparagus, smooth and of an indefinite
dark color with bronze lights. It was growing, apparently, on the
bark of fallen trunks and large dead branches. It is clearly a type of
parasitic tiril. Not being able for obvious reasons to touch the plants,
I made a few drawings which I've sent to Jack for publication.
These tirils look perfectly normal to me, but my dear Metzen, how
on earth are we to explain their presence on what is the biodegrad-
able base *par excellence,* a trunk that in a tropical forest is bound to
be rotted by damp and insect damage within a matter of months?
What happens when a parallel plant, by nature indifferent to the
passing of time, entrusts its destiny to such precarious conditions?
The credibility of all our scientic work over the last five years is in
jeopardy unless we can find a clear answer to this disturbing question."

On his return from Brazil, Scheinbach started a series of experiments to find plausible connections not only between the two botanies but between the kinds of time that in such totally different ways condition the existence of parallel plants and common plants. These experiments are still going on, but it is generally agreed that they have small hopes of success.

Tirillus odoratus

We know that the luster of parallel plants is often increased by the adherence to them of a kind of wax, a colorless substance known as emyphyllene. This substance is usually without scent, but exceptions to this rule occur in the case of certain varieties of *Giraluna* and in that of the *Tirillus odoratus* of the Mexican Sierra Madre. The Machole Indians think this property is responsible for the hallucinatory effects that have recently been studied by a team of neurologists led by Carlos Manchez.

T. odoratus is found in the mountains of the northern part of the Sierra Madre. It is a rare plant, especially as its habitat is among the exposed roots of *Olindus presistanus,* a tree well on its way to extinction throughout the American continent. The tiril is quite small, scarcely ever over twelve centimeters in height, and is almost black. As it is always found in shadow, it is practically invisible. In the first ten days of April the infinitesimally fine layer of emyphyllene deveolps a strong sour-sweet smell which has an aphrodisiac effect on the young Macholes, producing violent reactions which have now become completely ritualized.

The Macholes bear a close physical and cultural resemblance to the Huicholes, who live on the eastern slopes of the Sierra. It is thought that the two tribes were originally a single group and that they came to be divided by the misfortunes of war at the time of the Spanish Conquest. The Huicholes are known for the march of the *peyotero*s, an annual ritual the purpose of which is to gather peyote, a mushroom rich in mescaline. It is not improbable that the march of the Macholes to gather the *T. odoratus* has its origins in the famous march of the *peyoteros*, or vice versa; the latter, incidentally, is the coveted goal of certain foreign devotees, who arrive from all parts of the world during the short peyote season.

Some scholars are of the opinion that the behavior of the

PL. XII Parasitic tirils

Macholes is not due to emyphyllene at all, but rather to a phenomenon of *collective suggestion* arising from cultural motives that assign a symbolic and almost tantric value and meaning to the perfumed tiril. But Manchez, though admitting that the wax does not appear to contain any known hallucinatory substances, not only attributes the powerful aphrodisiac effect to it but also thinks that it provides the explanation of a particular conundrum facing students of this plant: I refer to the fact that the tirils allow themselves to be picked and carried away, even though during a very brief period each year.

Be this as it may, on the first day of April the Macholes from the scattered mountain communities, scarcely more than a hundred of them, gather in the village of Taichatlpec—which consists of a dozen wretched hovels roughly constructed of mud and cane—and there the march begins. The young women, who have also come in from the other settlements, retire into the great ceremonial hut known as the *cheptol*. This primitive gynaeceum, which is taboo for men, is usually reserved for menstruants, women about to deliver, and the young girls of the tribe for their initiation ceremonies. It stands about three hundred meters from the cluster of hovels that is Taichatlpec. For this occasion the young girls decorate the *cheptol* with tissue-paper streamers of many colors, while the old women, in a ceremony that goes on for most of the night, shut the door of the great hut and seal it from the outside with the gummy resin of the *manteca* (buttertree). The Machole women remain sealed up in the *cheptol* for three days, during the whole of the march of the *tirilleros*, and for all this time they neither sleep nor eat. They sing in low voices almost ceaselessly and anoint their bodies with the oily and pleasant-scented juice of the *tlac* fruit, a large earthenware jar of which stands in the middle of the hut, which otherwise is completely empty.

As soon as the first tiril is found in the shadow of an *Olindus*, the *tirilleros* begin a kind of march-*cum*-dance which gradually mounts to a frenetic rhythm and is kept up nonstop for three days and nights. As the plants are picked they are handed from one *tirillero* to another so that everyone can inhale the scent deeply. At midday on the second day, when sexual excitement has reached a peak of erotic hysteria, the men begin their journey back, still dancing, singing, and calling on the god Aptapetl.

Holding the tirils high above their heads, they gather in front of the *cheptol.* At a sign from the shaman, who alone has the divine authority to break the taboo, six *tirilleros* known as *"tlocoles"* hack down the frail rush door with machetes, and the horde of *tirilleros* pours into the *cheptol.* The orgy that follows may well last as long as the march, for two or three days. It has been described by John Meesters, perhaps the only traveler who has been present at this extraordinary event, though even he observed it only through a chink in the back wall of the *cheptol.*

Meesters arrived at Taichatlpec by chance, just as the *tlocoles* were on the point of breaking down the door of the *cheptol.* He was accompanied by a young Huichol Indian guide. Taking advantage of the excitement of the mob in front of the cabin, Meesters succeeded in tethering his horses in a nearby wood and hiding, with his young guide, at the back of the building. There they stayed for most of the night, then made off before dawn for fear of being discovered. Until sunset they could make out what was going on inside, if only rather vaguely, since there was no light except the sunlight that came in through the small doorway. Later on they were able only to hear the noise of body movements, groans, and every now and then a collective chant.

The young Huichol, who had studied at the Indian Affairs Center at Guadalajara, was later able to explain to the ethnologist the meaning of what he had witnessed. He told him that the Macholes have a generally quiet family and community life based, surprisingly enough, on monogamy but not on patriarchal power. The function of the orgy caused by the *Tirillus odoratus* is that of institutionalizing an occasion of total sexual promiscuity, and in such a way that the children conceived during those days are not attributable to a specific father. Instead, they grow up without the dominance of a father-owner and are lovingly cared for by the whole community. In the Machole tribe this system has eliminated not only the well-known neuroses due to father-son relationships, but also inheritance and private property. This has happened also among the Huicholes, but for quite different reasons.

Tirillus silvador

South of Lake Titicaca on the plateau of the western Cordilleras of Peru, where our own domestic potato first grew, there are a number of plants which thrive in spite of the altitude and are of prime importance to the diet of the Indians. Such are the oca (*Oxalis tuberosa*), the ullucu (*Ullucus tuberosus*), and the quinoa (*Chenopodium quinoa*). In the barren areas some hundreds of meters above the natural limits of these plants there are patches of tirils, of the variety *T. silvador,* which on account of their strange behavior have been attracting the attention of scientists for some years.

In general appearance these Andean tirils are not very different from other varieties recorded here and there in the remoter regions of the earth. Averaging some twenty centimeters in height, they live pressed together in dense colonies. Their color is that of the classic tiril: dark gray with bronze lights. They are without scent and have not changed their distribution or habitat within human memory. For some unknown atavistic reason the Indians do not touch them, and even the llamas passing near them on their way from one pasture to another are very careful not to step on them.

The extraordinary and so far unexplained feature of these little plants is that on clear nights in January and February some of them seem to emit shrill whistling sounds that are perfectly audible two or three hundred meters away. The sound is similar to that of the song cricket native to the Cordilleras, and many Indians believe that the crickets hide among the tirils. Their belief is strengthened by the fact that, like the cricket's song, the whistle of the tiril stops as soon as anyone approaches the place where it seems to be coming from. But experiments have shown that the cricket could not survive above 3,500 meters, while the tirils are located at 5,000 meters above sea level, or even more. José Torres Lasuego of the University of Chaluco, who has made a number of expeditions into the area, has studied the problem thoroughly without being able to give any proper explanation. Thousands of plants have been examined, but none has revealed any morphological anomaly that might in any way allow for the production of sound. The hypothesis of "draft strips," intermittent gusts of wind that cause certain strategically placed plants to vibrate, seemed at first to have some plausible basis, but careful experiments have shown it to be as unfounded as all others. Torres put forward the idea that certain individual plants were gifted with the ability to produce sound, and that this had the specific defensive function of keeping the llamas away. He did in fact notice that when a llama approaches the plants the whistling sound increases in pitch, while the animals, though skirting the very edge of a bed of tirils, appear to take great care not to step on them. However, it was later discovered that there was no connection between the two phenomena. The length of the sound wave of the whistle of *T. silvador* was found to be 12,000 melodin (A.S.I. system), while we know that the llama, like others of the camel family, cannot pick up vibrations above 9,000 melodin.

Torres's team installed a series of minimicrophones along the edge of several tiril beds and made many hours of recordings, which unfortunately have only documentary value. However, this capable Peruvian scientist is said to be now completing work on an instrument able to fix the exact point of emission of the mysterious whistling sounds. He hopes that this may lead to the identification of the plants responsible for the sounds, which is a necessary first step toward any adequate explanation of the phenomenon.

The Aymarà Indians, who are native to the Sierra, have numerous fables and legends in which these whistling tirils figure as heroes. One of these in particular is known throughout Peru in the form of a *corrido* (popular song), the origins of which go back to the Spanish Conquest. Until a few years ago the words of the song were attributed to Manuel Gonzales Prada (1848–99), the poet, essayist, and radical philosopher, but the musicologist José Manuel Segura has traced a manuscript dating from 1830, only six years after the Battle of Ayacucho (December 1824) in which Captain Sucre, with the aid of Bolívar, ensured the independence of Peru from the Spanish crown. The corrido tells the story of a patrol of Spanish soldiers, under the command of the notorious Captain Malegro, engaged in a search for Manolo Perchuc, a young Indian accused of having killed a dozen conquistadores with his *cachupote*. The young man had hidden in a cabin in the high Sierra, but he was betrayed by a jealous woman. One night the Spanish patrol advanced cautiously upon the cabin, which happened to be surrounded by patches of *T. silvador*. Suddenly, from every direction, came piercing whistles. The Spaniards thought they had been caught in an ambush and ran. Believing that the girl had betrayed them, they condemned her to death.

THE
WOODLAND TWEEZERS

The woodland tweezers, like the tirils, are social plants. They live in the shade of the Manengo trees in the jungles of Indonesia, under the *Kieselbäume* of the Black Forest and between the roots of the *ben* trees of Tetsugaharajima, in colonies which sometimes exceed a hundred individual plants. The most casual glance at these plants leaves no doubt as to how they got PL. XIII their name. Two winglike leaves, much like those of the *crassulae* species of Sugartongs, are symmetrically opposed in a gesture of rare elegance. With few exceptions, their slightly rounded tips always curl outward. Less basic than the tirils, the woodland tweezers are nevertheless fairly simple plants which in form have a good deal in common with normal plants growing in the underbrush, and it is only recently that botanists have become aware of their existence. As with many other parallel plants, they cannot be moved from one place to another. They turn to dust at the least contact with any object not native to their normal ecological surroundings, and only in a few cases has it proved possible to encase them in instant-drying polyephymerol. They are deep, dull black in color, rather like the tirils, and in the underbrush they are often hard to see against the shadows of the tree trunks. Sometimes this color is relieved by faint bronze lights due to the coat of wax that veils the outer surface of the leaves.

Professor Uchigaki Sutekichi, who holds the chair of sociobotany at Tokyo University, has published a number of papers on the supposed sociability of the woodland tweezers and the tirils, putting forward ideas that at first might appear arbitrary or even fantastic, but which are in fact worth serious consideration. Uchigaki thinks

that the distribution of the tweezers as we see it today is the final
result of an intricate series of maneuvers aimed at the conquest of
territory, and that these maneuvers bear the most extraordinary
resemblance to the moves of the game of Go. He says that the
woodland tweezers were originally sprouts from a complex rootstock
that was interwoven with the roots of the *ben* tree, a huge sper-
matophyte which grows only in the forests of the island of Tetsugaharajima. This rootstock was like a subterranean mind, planning
and storing the program for the gradual future distribution of the
shoots, putting the program into action, and controlling its various
phases. Strategic decisions which obeyed exact orthogenetic laws,
but which simulated a fierce struggle for survival, led in the end to a
kind of status quo, without winners or losers. Having abandoned the
illusion of self-determination and relinquished their fake weapons,
the individual plants ended by living a monotonous community life,
like aged veterans. It is in this last phase of the game that Uchigaki
thinks he can detect the moment of mutation to parallel botany.
The exhaustion of the energies which once promised to wrest from
time and space some existential meaning for the individual was consecrated in a collective immobilization. The fragile system that had
seemed to derive its ephemeral modes from the inevitably mistaken
answers to meaningless questions was thus replaced by the timeless
certainties which only a parallel condition could promise. Uchigaki
traces what he takes to be the history of a specific colony of tweezers,
drawing a parallel with a famous game of Go played many years
ago by the two celebrated champions Sharaku and Ugome.

Uchigaki's ingenious tour de force produced much excitement in
Japanese scientific circles, but did not fail to cause some perplexity
among his Western colleagues.

The game of Go was introduced into Japan by the Chinese legate
Yen Ta-yao during the Heiryū period, and more specifically in the
reign of the Emperor Shohei. It was soon taken up with enthusiasm
by the Japanese aristocracy, and in the course of time became the
Japanese national game. Go is thought to be the world's most
ancient game, the invention of which is usually credited to the
Chinese emperor Shu, who reigned at the beginning of the third
millennium before Christ. It is said that he thought it up to stimulate the mind of his eldest son Wang Wen, but it is a good deal more
likely that the old monarch, by studying the moves of the *ishi*, was
attempting to give a form to the precepts of the cult of Hsu-ch'uan,

PL. XIII Woodland tweezers

of which he was a devoted adherent. According to the Hsu-ch'uan doctrine, every action in life should consciously tend toward a moment of final stasis, the so-called *Ta-heng,* which in the ritual wrestling known as *Shou-hsi* represents the ultimate inextricable mutual immobilization of the wrestlers, the impasse at which the resignation of the potential winner is in perfect balance with that of the probable loser.

The *Ta-heng* can be arrived at by a process of meditation called *Chen-szu-liang,* in which thought is represented by a pendulum which swings from question to answer and from this, reformulated as a question, to a further answer or else to a paralyzing series of choices of priority in the everyday acts of life. It was to shed light on this second process, called *Ta-wang-hsi,* that the emperor Shu is supposed to have invented the rules of Go, in which Uchigaki thinks he has been able to detect the sociobotanical structure of the colonies of woodland tweezers.

The game in fact consists of a number of strategic choices of position (*moku*), the aim of which is to conquer territory on the Go board and to prevent one's opponent from doing so. With incredible determination and patience, Uchigaki compared the arrangement of the tweezers found by Sugino Kinichi a few years previously in the forest of Owari with the final position, the *Ta-heng,* in a game of Go, and by working backward went on to establish every move in the game. Thus following the course of the game he was able to trace the sequence in which the plants of the group had sprouted and to demonstrate that it had obeyed precise rules which were very similar to those of Go and had a development which, as he had suspected, was practically identical with the memorable match played by Sharaku and Ugome (Figs. 15 and 16).

Uchigaki's insistence on the comparison between the distribution of the shoots in the Owari group and the moves in a particular game of Go clearly shows leanings toward mystical and aesthetic aspirations which are quite alien to the Western scientific mind. All the same, we are bound to admit that the idea of a generating rootstock that creates and executes genetic programs has at least the merit of originality. Further weight is added to the theory by the consideration, mentioned by Uchigaki, that because plants cannot move around like animals their choice of location is always definitive and must be the result of a strategy in which all individual struggles have collective immobility as their final aim.

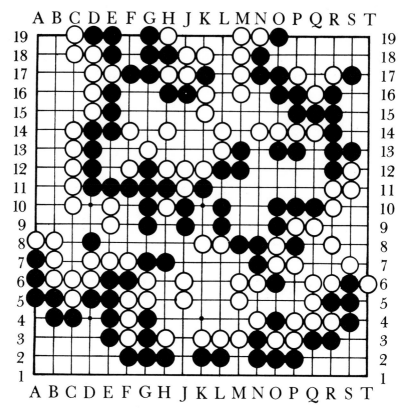

Fig. 15 The *Ta-heng* position in a game of Go played by Sharaku and Ugome

A bronze facsimile of the Owari group of woodland tweezers may be seen in the Imperial Natural History Museum in Tokyo. It forms the centerpiece in the largest room in the new wing entirely devoted to parallel botany. Around it are eight *go ban* made of *icho* (*Salisburia adiantifolia*): low tables with legs fashioned in imitation of the fruit of the *kuchinashi*. Now, *kuchinashi* is the name for *Gardenia floribunda*, but in Japanese it also means "mouthless," and is a warning to the spectators to remain silent while the games of Go are in progress.

It is in the Tweezers Room in the Museum that the National Go Championships have been played for the last few years, and it is by no means rare to see players, dressed in the traditional kimono, repeating move for move the game that was played thousands of years ago under the giant *ben* tree, and that in a final botanical *Ta-heng* was miraculously preserved intact down to our own times.

Fig. 16 The distribution of woodland tweezers at Owari

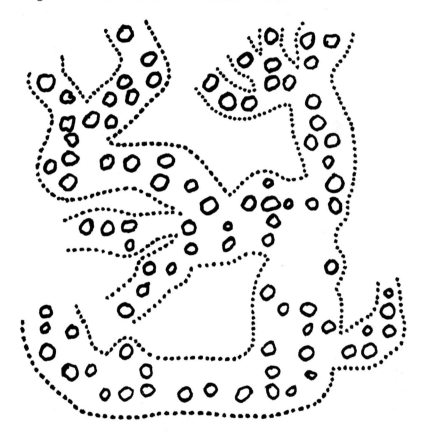

THE TUBOLARA

I have often mentioned the matterlessness of parallel plants, drawing attention not only to their entire lack of organs but also to the fact that they have no real interior. Oskar Halbstein extends this notion, typical of parallel botany, to everything in the world, observing that the interior of material objects is nothing but a mental image, an idea. He points out that when we cut something in two we do not reveal its interior, as we set out to do, but rather two visible exteriors which did not exist before. Repeating the action an infinite number of times, we would merely produce an endless series of new exteriors. For Halbstein the inside of things does not exist. It is a theoretical construct, a hypothesis which we are forbidden to verify.

The interior of parallel plants, moreover, eludes even theoretical definition. As we are concerned with a substance that is totally "other," that cannot be found in nature, it is literally unthinkable. Halbstein speaks of it as being of a "blind color," but to me it seems arbitrary and scientifically risky to draw even the most openly poetic comparisons with the normal world.

PL. XIV The *Tubolara*, which for the most part are found on the Central Plateau of Talistan in India, put the problem of the interior of parallel plants in a new and rather different light. It concerns not so much their matterlessness as their form, not so much the solid interior of their ambiguous substance as the hollow exterior— which, in a sense, is the external limit of the interior of the plant.

Here then is the paradox of the *Tubolara:* two interiors, one of which in normal terms would be its substance and which at bottom is responsible for its presence, is imperceptible, while the one we

PL. XIV *Tubolara*

would normally be inclined to think of as nonexistent, the void contained by the plant, is visible. The paradox is even greater when we think that the void within the tube, the visible interior, has a very precise function: that of containing, like a fragment of its own habitat, part of the environment in which the plant itself is contained.

In the Tantra the *Tubolara* represents the coexistence in time and space of the feminine and masculine principles.[1] It is *lingam* (male organ) and *yoni* (vulva), and as the physical union of man and woman it symbolizes the essence of created things. In the temples of Talistan, along with the statues of Krishna and Vishnu, we often see stylized representations of the *Tubolara* carved out of stone. Before the great temple of Shalampur there is one that is seven meters high, and as no one had ever been able to look at it from above it had for many years been taken for a *lingam*. It was only recently that some workmen who had climbed up to repair the temple roof were amazed to find that the pillar was hollow. As a result of this discovery, the ceremony of *Kala-tà*, which takes place before the sowing of the seed and used to culminate in the decoration of the supposed *lingam*, has had to be radically altered. At the end of the *puja* the officiating guru now climbs a shaky bamboo ladder and drops rice and flowers into the immense stone tube, the biggest *Tubolara* in the world.

In the jungles of Tampur in northern Rajastan there is a variety of *Tubolara* considerably smaller than that of Talistan. For the local inhabitants it is taboo. Ramesh Sashtra thinks that this taboo originated in the belief that the darkness inside the plants represents the dark night of death, populated by white bats which seize the bodies of the dead in order to devour their intestines. This talented Indian ethnologist supports his theory by pointing out that the *Tubolara* of Tampur are favorite haunts of *Elidoptera fenestralis*, a tiny white moth which can be vaguely seen fluttering in the darkness of the plants, transforming them into an "other world" that is no less threatening for being so small.

THE CAMPORANA

T he *Camporana* or kite-plant is known in different parts of
the world under a variety of names. In Ecuador it is called
cuavenco, like the kite-hero of the Aymol legend. In Haiti it
is the leaf *taihaque,* a name of African origin. In Dahomey and the
Upper Volta, where it must have grown in great profusion until the
vast migrations of the Okuna (at which time it became parallel on
PL. XV the very verge of extinction), the *Camporana* is known in Fon and
Korumba mythology by the name of *tialé.*

The *Camporana* is a monofoliate plant consisting of two main
parts: the sheathed *fustis* and the limb. Two varieties are known,
C. erecta and *C. reclinatus.* The former stand upright in the earth,
supported by the sheath around the long stem, while the latter lie
on the ground like a fallen leaf. In both types the central vein of the
leaf is simply a continuation of the stem, a large pedicel which
penetrates the limb, divides into two or three branches, and tapers
away to nothing toward the edge of the leaf. In the case of *C. reclinatus* it is especially pronounced where the leaf rests on the
earth, often presenting protuberances known as "root bosses" which
are, in fact, probably rudimentary roots. The limb of the *Camporana* is usually reniform and is small in comparison with the
stem. From some distance away, the plant sometimes reminds us of
the ceremonial fans (*flabella*) which used to be placed on either
side of the thrones of Eastern potentates and that even today flank
the Pope's gestatorial chair. The leaf itself is thin only at the edges,
which are often indented. It swells to considerable thickness toward
the middle, but thins down again in the immediate vicinity of the
central vein. James Forbes is of the opinion that before its parallel-

PL. XV *Camporana*

ization the swelling of the *Camporana* leaf contained a complete system of internal veins. According to this English ethnologist, to whom we owe much of our knowledge of the *Camporana,* the small excrescences visible on the leaf bear witness to an unsuccessful attempt by the vein system to become external.

The overall height of the *Camporana* varies from 35 to 190 centimeters, but specimens of *C. reclinatus* seldom exceed a meter. During an expedition in the Upper Volta, Forbes succeeded in photographing one 135 centimeters high in which the outlines of a face had been cut. The plant was at least seven hundred years old, and from the local Korumba tribe Forbes heard several legends that appeared to confirm his hypothesis that it exercised a totemistic function.

Of particular interest is the cosmogenetic myth of the *tialé* which tells of a tree that grew near the source of the River Dwon. This tree had thirty-seven huge leaves, and belonged to the Leopard. Now there was nothing more precious in the whole world, for the Leopard had only to touch one of the leaves with his tongue and whatever he wished for would appear by magic. Over the years the Leopard had wished for Moibu, the snake who is coiled round all the islands, and the clouds that give rain, and the turtles of the earth, and even Keple the great spider. But the Leopard was old, and one day he felt death approaching. What was he to do with his tree? He went to ask advice from Tok the Hare, who could talk better than other animals because of the split in his lip. "What should I do with the *tialé?*" he asked. Tok advised the Leopard to give it to his sons before he died. But the Leopard asked: "How can I divide the leaves between my twelve sons?" "Kill one of your sons, or two or three," replied Tok, "until you can divide the leaves equally among those that remain." But the Leopard realized that this was not possible: he did not succeed in dividing the thirty-seven leaves by any number whatsoever. He asked advice from the Ant, but the Ant could not find an answer. He asked the Turtle, and Twembo the Serpent, and even the God Nawaki, but none of them could find an answer. Then the Leopard went to Tsò the Caterpillar, and Tsò said: "Take me to the tree *tialé.*" He climbed onto the Leopard's tail and was carried there. He set to work at once and ate one leaf.[1] Now the Leopard could give three leaves to each of his sons and die in peace. Tsò the Caterpillar went down to the River Dwon and there he excreted the remains of the *tialé* leaf. When the river saw the excretion it licked it with a wavelet, and lo, in the midst of the waters there was a canoe, and in the canoe were Kwep and Lamu, and from them sprang all the sons of the Korumba people.

James Forbes, whose great love is botany but who is by profession an ethnologist, has made an interesting observation on this myth, which is the origin of the totem *tialé*. "It is no mere coincidence," he says, "that the myth is based on a prime number, which is a fairly sophisticated concept. Of all African peoples, the Korumba are the best versed in arithmetic, as can be seen from their game of *twam-ha-rà*, which is played with a hundred and eight pebbles, each with a different numerical value."

African mythology has survived in the voodoo cults of the black tribes of Central America and Brazil. In Haiti and elsewhere in the West Indies, where the myths of the Ivory Coast have been preserved with greater purity, the tree *taihaque* figures as the divine instrument for the creation of all living things. According to legend, this immense tree stood on the cloud called Waikò and bore fruit containing the plants and animals, the fishes and the birds, the turtles, and also man and woman. When the God Nyambe had finished making the great ocean and all the islands that float in it, according to the Haitian legend, he shook the *taihaque* with mighty strength and the ripe fruits fell to earth and split open, setting free the creatures who had matured inside them. When all the fruits had fallen, Nyambe took it into his head to send the angels to inspect the results of his work. He put each of them on a leaf and shook the tree once more. The leaves fluttered earthward like so many butterflies. The angels made sure that all was well, but when they were about to return into the heavens men begged Nyambe to let the angels stay with them. The God allowed the twelve angels to remain on earth and watch over the happiness of men. The angels planted their leaves stem downwards in the earth, to keep them alive in case Nyambe should ever recall them to heaven. When white men came with guns and forced the Wakonga people into slavery, Nyambe recalled the angels to his side. On Mount Wabika in Dahomey there are twelve holes: they were left by the angels when they tore up the *taihaque* leaves from the ground.

In his book on African myths published more than thirty years before the discovery of parallel botany, Vobenius sees the *taihaque* leaf as a huge monofoliar plant which actually existed in Dahomey and probably became extinct in the middle of the seventeenth century, at the time of the worst incursions of the French slave-traders.

He could not have known the true nature of the plant, but he
guessed at its exceptional importance, and thought it possible that

Fig. 17 The *behin* of Fra Girolamo di Gusmè, from
a contemporary print

the voodoo version of the ancient myth represented, in its conclusion, the destruction of the African cult objects by the French.
Uncertain whether to accept the hypothesis of a real plant or that of
a cult object, the great German ethnologist unknowingly touched
the fringes of the new botany and even appeared, though only in an
ethnological context, to recognize the *Camporana africana.*

The distribution of the "root bosses" on the leaf of the *Camporana* was studied by medieval alchemists and sorcerers. It is from
their writings, which are sometimes extremely enigmatic, that we
learn of the existence of the plant in not so remote times. In the
Ziharmi we read: "All things come to pass down there as they do up
here. The figures formed by stars and planets reveal things hidden
and mysteries most profound. In the selfsame way on the skin of
the *behin* leaf there are bosses which are the stars of the plant." The
sorcerers examined thousands and thousands of leaves in the hope
of prophetic signs, and wandered endlessly through the forests in
search of the legendary leaf. Fra Girolamo di Gusmè, alchemist
and teacher of celestial sciences, left us a series of diagrams illus-
trating the spots of the leopard and the distribution of moles on the
human body. Among them, however, is a drawing of a leaf that is
without the least doubt a *Camporana* (Fig. 17). In it the position of
the bosses is shown in relation to the heavens and to the signs of the
Zodiac. "He who finds the *behin* may learn his own fate," declares
Fra Girolamo, adding that according to whether the bosses are on
the right or the left of the leaf they refer to good and lucky events or
to bad and unlucky ones. What the worthy Friar neglected to tell us
is which is the front of the leaf and which is the back, so that we
can never be sure of knowing the right from the left. A further
complication is that when the woodcuts were printed from the
original drawings they were in all probability reversed, though we
are not one hundred percent certain of this. It therefore comes
about that in the unlikely event of anyone stumbling across a *behin*
leaf in some distant forest, we can at least be sure that knowledge
of his fate and future would fortunately be denied him.

In the Jardins Publiques of Ouagadougou, capital of the one-time
French colony of Upper Volta, there is a marvelous little octagonal
cast-iron conservatory attributed to Eiffel. The decoration was added
later, during the *art nouveau* period, and the intricate floral motifs
so typical of this style interweave with the real plants visible
through the dusty panes of glass.

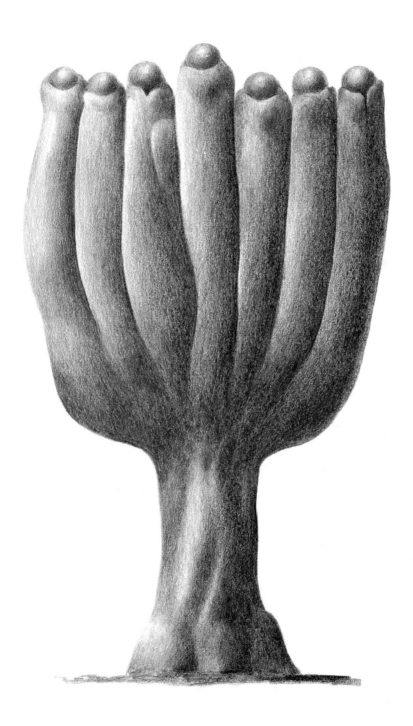

PL. XVI *Camporana menorea*

In this conservatory are all sorts of plants typical of the hinterland of the Ivory Coast, and especially those which grow along the three upper branches of the River Volta—the Black, the Red and White, and the Oti. In one corner there is a group of bronzes donated to the colony in 1908 by Jean Philippe Audois, governor of the territories which now form the state of Togo. Audois, who like many French bureaucrats of the time was also a naturalist and a fairly good minor poet, wrote of the "anciennes audaces de plantes solitaires / nègres botaniques d'herbaires silencieux / noyés dans le temps d'un fleuve phantôme." The words *ancient, solitary, black, silent,* and *phantom* form a verbal chain which leaves no doubt that the plants he was describing were parallel plants which centuries before had flourished along the banks of the great African river.

The bronzes on show in the conservatory are nine in number. Five of them represent *Camporana* found in the region, known to the indigenous tribes as *tialé* or *keletià*. Of the other four, three are plants belonging to normal botany while the fourth cannot with certainty be assigned to either realm. Some experts think that this fourth plant is a *Sigurya,* but the specimen does not have a sufficient number of pendulants to justify this hypothesis. At the same time the plant does present a number of features which would be decidedly abnormal in the general run of tropical plants. The five *Camporana* represent all the erect varieties known to us. Two of them are nearly two meters tall, and have a rather irregular distribution of the bosses. One of them is simply a miniature variety of the larger ones, with almost identical proportions. Of the remaining two bronzes one represents *C. menorea,* the name of which derives PL. XVI partly from its smaller size (Lat. *minus* less) and partly from its marked resemblance to the menorah, the seven-branched candlestick used in Jewish worship (the name was in fact conferred on it by the Israeli naturalist Ismael Brodsky).[2] The other bronze is a typical specimen of the small *Camporana* "For Grace Received," thus baptized by the sisters of the Tuogoho Mission, on account of its approximate heart shape. The good sisters, who did not think twice about compromising with the animistic cults of the local population, invented a story which culminated in a miracle worked by San Trino of Montassano, whose prayers succeeded in changing a poisonous plant responsible for the deaths of thirty children—the terrible Fuahamec[3]—into a harmless plant which by dint of the miracle took on the form of the Sacred Heart of Jesus.

THE PROTORBIS

Above an Empire *chaise longue* in the Gilded Room of the Château Nouilly at Vincennes there is a large painting by Gérard Méliès, a hack painter, cousin of the great filmmaker, who enjoyed a certain notoriety in Parisian artistic circles at the end of the last century. It is a portrait of the grandmother of the present Countess Amandine. While the figure of the lady herself is painted realistically, in a manner faintly reminiscent of the style of David, the background reveals a temperament which perhaps at certain other times in the history of taste might have found expression in a certain lyrical élan. It shows a vast sweep of landscape, with bleak mountains huddled fearfully beneath the gathering storm clouds. In the narrow valleys there are black cypresses, while here and there on the rugged horizon is the silhouette of a distorted oak tree.

But in fact this is not the landscape it appears to be at first sight. Instead it is a still life, a heap of unusual plant shapes, forms intermediate between a mushroom and a potato, from which sprout a few leaves like those of parsley or celery.

The type of ambiguous play, the gambit that enables the spectator to transform the landscape into a still life and vice versa, takes one completely by surprise: it reveals that hereditary stroke of genius that like a leaf of an unimagined color buds forth from time to time on the family tree of the Méliès'. But if the importance of this extraordinary picture stopped here we might be justified in taking it for the mere whim of a gifted but somewhat bizarre temperament. Its importance lies elsewhere, and it is only recently that we have been able to perceive its true extent.

The mushrooms lying scattered on the outspread green cloth,

which in the "landscape" we see as an undulating green meadow in the midst of bleak mountains, are in fact a number of *Protorbis foetida* which Méliès, a restless soul and tireless traveler, brought back from Asia Minor where he had gone on an expedition with his biologist friend Jean Entigas. Méliès was a man of revolutionary tendencies who did not hesitate to accept commissions from the richer echelons of the Parisian *bourgeoisie,* people who while not wishing in the least to accept any ideological discipline were fond of adorning their evenings with some of the more eccentric members of the intellectual élite. It was perhaps to unburden a certain sense of guilt that the painter invented this *trompe l'œil,* being absolutely certain that the plants were so rare that no one would ever realize his act of provocation, which in any case was in poor taste and of small political significance.

And so things stood until 1965, when Professor Pierre-Paul Dumasque, a childhood friend of the Nouilly family, recognized that fantastic landscape as an important group of parallel plants. The circumstances leading up to the discovery of these *Protorbis* in Persia are not altogether clear, nor do we know why they remained in the possession of Méliès. The fact that Jean Entigas was a keen collector of his friend's work suggests that the mysterious tubers might have been given to the painter in exchange for one of his pictures. We have since learned that Méliès kept them in a transparent case, like a fish tank covered with a sheet of glass, along with a thousand other useless trifles he had brought back from his travels, and which lay piled and stacked on the shelves and all other available surfaces in his studio.

After the death of Méliès his sister Melinde inherited all this bric-a-brac. Dumasque, spurred on by his discovery of the meaning of the picture, made tireless investigations which led him eventually to a son of Melinde's, a doctor in Arbières (Indre-et-Loire) who had in his turn inherited this weird collection. It was not hard to persuade the doctor to donate the fish tank and its mysterious inhabitants, which in all these years had not shown the least sign of deterioration, to the Parallel Botany Laboratory of the Jardin des Plantes in Paris, where both Dumasque and Gismonde Pascain were able to study them at their leisure. The results of their researches were later published in a special issue of *The Journal of Parallel Botany* (October 1974) under the rather gimmicky title, "*Protorbis*—a parallel mushroom?"

If we are today in possession of those basic items of knowledge that enable us to carry on our study of parallel phenomena, we owe this in good part to the chance discovery of this picture by Méliès. But more important in the long run was the wise and patient research which enabled the two French scientists to define the true nature of *Protorbis,* which is indeed anomalous and bizarre in the extreme.

PL. XVII The *Protorbis,* of which *P. foetida* is only one variety, has doubtless some points of resemblance to the mushroom family. These include form, color, and opacity. What distinguishes *Protorbis* from its cousins on the other side of the hedge is the irregularity of its outlines and the very massiveness of the testula which is less like the cap of a mushroom than like some vast black truffle. The importance of *Protorbis* lies in its lack of precise dimensions. It can be of any size, from the infinitely small to the infinitely large, a respect in which it reminds us of the first flora ever to exist, which in its perfect transparency was absolutely invisible and therefore not subject in any way to the concept of dimension. It is generally thought, in fact, that *Protorbis,* along with the tiril, is among the earliest of parallel plants. Certain specimens in the deserts of New Mexico and Arizona, referred to by Entigas, are as big as the nearby mesas, and are indeed often mistaken for these hills, with their flat tops, in spite of the differences in form and matter. *Protorbis* is in fact composed of a substance which has only superficially the aspect of stone. If it is struck with a normal geological hammer it emits a high-pitched metallic sound totally at variance with its heavy and opaque appearance. The matterlessness which is attributed to the greater number of parallel plants must in the case of *Protorbis* be seen in a different light and completely redefined. In the sense of a material without any verifiable interior, of regular density and lacking any measurable specific gravity, we can still speak of matterlessness or nonsubstantiality. But at the same time anyone not versed in the ways of parallel botany might see or touch the plant and pronounce it—according to its size—to be a large hill or a virtually shapeless metal object.

Apart from *P. minor,* which disintegrates instantly at the least touch of a hand into the merest pinch of white powder, all specimens of *Protorbis* may be transported (size permitting), while their conservation requires no special techniques or environmental conditions.

Dumasque lists seven varieties of *Protorbis,* divided chiefly by

PL. XVII *Protorbis*

difference in size. These are: the Colorado *Protorbis; P. foetida; P. minor;* the Katachek *Protorbis; P. bisecta; P. inopsa;* and *P. torbis.* This classification, which is now accepted by all scholars, was initially much criticized. And it seems to me that the critics had a good point. How can a plant which has no fixed dimensions, for which any size at all is theoretically possible, be divided into seven varieties by a criterion based primarily on size? Except for *P. minor,* a special case owing to its unique properties, this arbitrary division seems to conceal the most salient feature of the species: the way it varies from one plant to another, so that each individual is practically a species of its own, a never-to-be-repeated freak.

Furthermore the shape of the *Protorbis* is less constant and conforming than that of other plants, a fact which in itself makes a general description more difficult. Each plant has not only its own individual size, but also a shape of its own. Some of the *Protorbis* of Colorado and New Mexico, which stand in the desert shoulder to shoulder with the mesas, have trunks nearly as broad as the testula itself. In shape they are virtually cylindrical, and quite indistinguishable from the surrounding hills. In other cases, such as the *Protorbis* of Kamanchistan, discovered by Kowolski's son, the trunk is no bigger than that of an oak.

In nearly all known specimens the testula is rounded and smooth. It is in fact rather like the top of a human skull, or reminiscent of the cap of a mushroom except in respect to its size and color. In the tundra of Katachek, in the course of a journey through Siberia and in the vicinity of the Chinese frontier, Dumasque came across a specimen of *Protorbis* with a testula covered with protuberances similar to tirils. At first sight he took these for parasitic growths, but closer examination revealed that they were integral parts of the plant, "almost as if by this morphological anomaly it wished to voice a protest against its leaden immobility, a rebellion against its own amorphous and brutish appearance, by making a desperate attempt at flight toward the lightness and elegance of aerial things."[1] Now known as the Katachek *Protorbis,* it has PL. XVIII entered the catalogue of parallel botany as an anomalous variety, but it is not impossible that it is the only known example of a different species altogether, with only a few tenuous analogies with *Protorbis.*

The Indian *P. minor,* which exists in relative abundance in the jungles of Jandur and on the Tampala Mountains, is no bigger than a mushroom. It was the Middleton expedition which first used steophytirol to encase these plants, which are usually untouchable. Those they brought back are now on display in their little plastic cubes in the Parallel Botany Room of the Birmingham Natural History Museum. Some might express surprise that Birmingham, one of the major industrial centers of Great Britain but certainly no intellectual mecca, has one of the most important and complete collections of parallel plants in the world. However, we must bear in mind that the coal beds of the area long ago made Birmingham a great steel town, and it is only logical that economic interest in the natural resources that lay just under the surface of the earth should involve those sciences which were also developing dramatically during the eighteenth and nineteenth centuries and were closely linked with the impetuous advance of technology during the Industrial Revolution. The discovery of important beds of fossils encouraged paleontological research in the region, and these in turn attracted zoologists, botanists, and, ultimately, students of parabotany such as Wells and Joseph Middleton. The cultural development of the city, of course, did not stop there, and Birmingham acquired a great university, municipal art gallery, and symphony orchestra.

Even in early days a number of rich industrialists became aware that their economic wonders, performed at the expense of the working (or rather, toiling) classes, were causing increasing resentment, and they realized that among the intellectual élite of the country their persons were looked upon with no great sympathy. They therefore sought for ways to attach their names to enterprises of high cultural and moral prestige which at the same time were not too remote from their immediate sphere of interest.

The idea of an important museum of natural history specializing in the subterranean sciences was suggested in 1896 by Sir Oswald Otterton at a historic meeting of the Carbon Club, and a few years later it was an established fact. Generous bequests were remembered by a supposedly grateful public on the commemorative plaques that line the entrances to each room. The Parallel Botany Room was originally financed by Sir Jonathan Hoverley, and he is duly immortalized on a vast sheet of black marble.

The room is appropriately grand, measuring about ten meters by five; on one side are two great windows overlooking the park, where trees of every known species, each bearing its name and title, are frozen as if on a grass-covered stage in the midst of a botanical

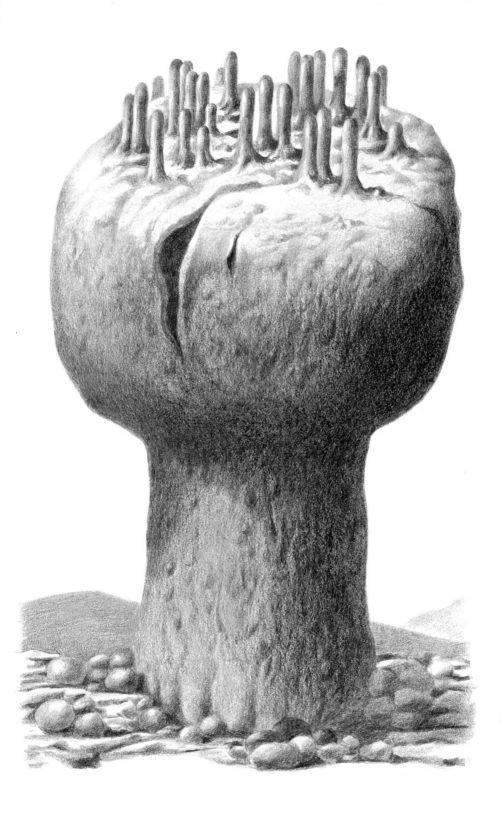

PL. XVIII The Katachek *Protorbis*

comedy by Samuel Beckett. In comparison, the parallel plants arranged along the other walls and in the three central showcases seem, by some magical chance, to have found their natural environment. Perhaps this is because the context is so obviously that of a museum, slightly outside of time and isolated from the vulgar bustle of the world. Outstanding in size are the great *Camporana* leaf, three meters high, which dominates the room from the center of the long wall; the model of *Giraluna gigas,* a bronze copy of the smallest plant in the Lady Isobel Middleton group, which was donated to the museum by Maessens, who directed the reconstruction of the group (now in the British Museum); and the three so-called trunk *Solea,* on loan from the Laboratorio delle Campora.

The showcase nearest the entrance contains a heterogeneous collection of important documents, including the famous letter of Jacopo della Barcaccia (donated by the Italian government), Malgueña's notes on the *Sigurya,* the camera with the polyephymerol lens that enabled Norton to photograph the Tampala *Giraluna,* and a small sculpture by Arp which is identical with the plaster model of the *Artisia* on display beside it. The showcase at the far end contains fragments of fossils, concretions and impressions, as well as a small collection of seeds, fruits, fragments of pseudo-bark and other paramimetic objects donated by Sir John Everston.

The central showcase, larger than the others, houses no less a treasure than the *Protorbis minor* from the Middleton expedition. It is equipped with a special lighting system designed to display the three-dimensionality of the plants, for in normal lighting, stuck in their plastic cubes, they tend to lose that appearance of weightiness characteristic of all the *Protorbis* varieties. The collection consists of twelve very similar specimens, nearly all in a perfect state of inclusion.

A particularly beautiful specimen is the one which Lady Isobel Middleton christened "Beginner's Luck," because it was the first she found. The overall height is eight centimeters and the testula is rather irregular, with a diameter roughly equal to the height of the plant. Its black color is definitely blacker than that of any of the other specimens in the showcase. The plant gives the impression of hugging blackness to it, like the night which once hid it and which still clings to it like an opaque skin.

This indefinable blackness of the *Protorbis,* and especially of the Birmingham *P. minor,* provided one of the most interesting and

original features of a lecture given by Norton at the Carbon Club
and published in part in the annual *Proceedings* of the Club. For
Norton, the black of *P. minor* is the *ne plus ultra* of the color of
parallel plants. To perceive it in its fullness and to glimpse the
significance of it we must look closely at our whole relationship with
the inside of things. "We deny," said Norton, "that there is a difference between internal and external landscape, and we tend to
transfer to the inside of things their solar skin, just as it is, without
change, absurdly illuminated by a nonexistent light. Thus we imag-
ine the inside of our bodies: a many-colored landscape in which
blood-red and bile-green mingle on a palette that is only too familiar. We have only to suspect that there might be a black organ to
realize that Satan has possessed us and that only the most strenu-
ous exorcism will ever bring the light of day back into our vitals.
While anyone who sees the inside of things as an impenetrable
darkness runs the risk of being diagnosed as an incurable depres-
sive, if anything, our real neurosis consists in our lack of ability to
see and accept things as they actually are."

Norton goes on to tell how when he was in India he learned to
think of the inside of things without doing violence to their integrity, just as he managed to think of his own most precious organs,
lungs, heart and liver, as black flesh immersed in the utter blackness of his body, and to derive no feelings of discomfort from the
thought. And then, one day, he suddenly understood the blackness
of *Protorbis*. "It is logical enough," he wrote, "that plants which
have no real and proper internal substance, but only an existential
continuum circumscribed by its own formal exhaustion, should not
have an exterior like that of the other things of the earth. The
outside which we see in their case is not a kind of wrapper that
contains, conceals and protects nonexistent lights and colors, but
merely the visible limit of their internal darkness. They present
themselves to us in all their utter nakedness, showing us exactly
what they really are."

The Amished people of Tampala are perfectly familiar with
P. minor, which they call *bahan*. An English officer stationed in the
area not long after the First World War, a certain Major James
Ronaldson, became interested in their ethnological problems and
has left us an enlightening document on a number of local legends
in which the *bahan* figures explicitly. In those days no European
had ever seen the plant, and even Ronaldson was convinced that it

was an imaginary plant, the fruit of the folk imagination. But many years later he happened to stumble across the *Proceedings* of the Carbon Club, and had no difficulty in identifying the plant described by Norton as the legendary *bahan*. Although by then over eighty, Major Ronaldson got in touch with the well-known botanist-photographer and sent him his account of the legends of Tampala. The two later met at Bensington in Kent, and fragments of the conversation that took place in the cottage garden under the great weeping willow which Major Ronaldson, an inveterate punster, liked to refer to as his "weeping widow," were published as a long appendix to the *Annals* of the Birmingham Museum of Natural History in 1974.

Of the legends recounted by Ronaldson, the most interesting is the one most explicitly concerned with *P. minor*. Here it is in full:

NANDI AND THE NIGHT

Every spring Lord Krishna used to come down from the mountains to graze his calf Nandi[2] in the green meadows of the Andrapati valley. But one day, though the sun shone hot in the sky, he found the meadows still covered with snow. When night fell Nandi wept and said to Krishna: "My lord, I am hungry. Make the snow melt and the grass grow so that I may eat, and grow strong, and be happy."

So Krishna went to visit the Night, and said: "Nandi is hungry. Make the snow melt and the grass grow." But the Night answered: "Krishna, I am only the Night. I cannot melt the snow." Then Krishna said, "Tell the sun to melt the snow." But the Night answered, "I am only the Night. I cannot command the sun." When Krishna heard these words he grew angry and said, "Then I will take a piece of you to feed Nandi, who is hungry." And so he did. He brandished his great sword in the sky and a piece of the Night fell broken at his feet. Lord Krishna gathered up the bits and took them to Nandi. Nandi ate what Krishna had brought him, and when he was satisfied looked up at the sky and said, "Lord Krishna, you have made a hole in the heavens." And Krishna answered and told him, "It is the moon." So Nandi slept. When he awoke at dawn the snow had melted and the fields were green. For three days

Krishna grazed his calf in the valley. Then the bird Vardatur came and carried Krishna and Nandi away. The crumbs from the piece of the Night remained, scattered under the great genensa trees. They are called "bahan" and they cannot be touched, for they are the food of Nandi the sacred calf of Krishna. If they chance to be touched by hand they will turn back into sky and fill the hole in the Night, and in this way the moon will disappear forever.

This legend is the clearest proof that the *bahan* of the Amished tribesmen is nothing other than the *Protorbis minor* discovered by Lady Middleton. The mention of the valley of the Andrapati and the forest of *genensa* trees, the description of the black-as-night color of the plants and their curious refusal to be touched, all these leave us in absolutely no doubt about it. It remains for us to find out when the legend originated. Ramesh Drapavati, Professor of Sanskrit at the University of Baroda, and a specialist in the *Vadrahana*, attributes the story in its present form to the Pachinah period, but does not rule out that it might be much older, perhaps even dating from the age of Akda. Such an attribution would confirm the hypothesis put forward by Maessens, according to which the *Protorbis*, along with the humble tiril, is among the first parallel plants of the earth.

The compact rudimentary form of both these species displays a fairly low level of plantness which nevertheless, in its bewildering immobility charged with frustrated violence, formed a prelude to the vegetable kingdom that thousands of years later was to alight softly on the black soil of our consciousness.

together. Inside these nests the tunnels and "halls" formed a topology of the greatest intricacy. In Mali there still exist several groups of these forts, now inhabited by wasps of the family Aligastorae. Because of the rounded rooftops which can be seen above the trees they are often mistaken for Dogon villages. Hard as stone, they have resisted the ravages of time. The builder ants, in fact, had a gland which secreted a gummy liquid, known as cementine, which on contact with the silicates of the earth produced a cement-like substance of great cohesion, practically indestructible.

The copulator ants (Fig. 18a) were similar to the builders, but were without even the most rudimentary organs of sight. However, they were equipped with sexual organs capable of an uninterrupted flow of spermatazoa. They lived in round chambers with slightly "vaulted" roofs in the "halls" of the nest, together with the queen ants (Fig. 18a), of which there could be as many as a thousand for each community. Continually stimulated and fertilized by the copulator ants, the queens alternated between copulation and the laying of eggs, which in the course of a single day could run into millions.

In proportion to the rest of their bodies the queens had an enormous abdomen which, like those of various termites, often reached a length of some thirty centimeters and a diameter of five centimeters,[1] and frequently lay coiled around the walls of the "halls." When emptied of eggs, this enormous organ partly shrank, leaving a long tube capable of peristaltic movement which conducted the sperm of the copulators to the inside of the reproductive apparatus properly so called.

But the most interesting caste among the antaphids was surely the eater ants (Fig. 18b). They were equipped with incredibly strong mandibles capable of chewing the vegetable matter on which they fed at a speed unequaled anywhere in the animal kingdom. They had two digestive systems, one of which was normal, of modest size and complexity, for their own nutrition, while a second, lateral system had the function of transforming the original nutritive substances into others readily absorbed by the builders, copulators, and queens. The eater ants were particularly fond of the tender fat leaves of the *Labirintiana* so that the monstrous voracity and prodigious increase in population of these insects put the survival of the plant in serious jeopardy. This led to the ingenious and rather quick mutation of the veining on the huge leaves of the *Labirintiana*, which thus changed its normal bilateral symmetry to

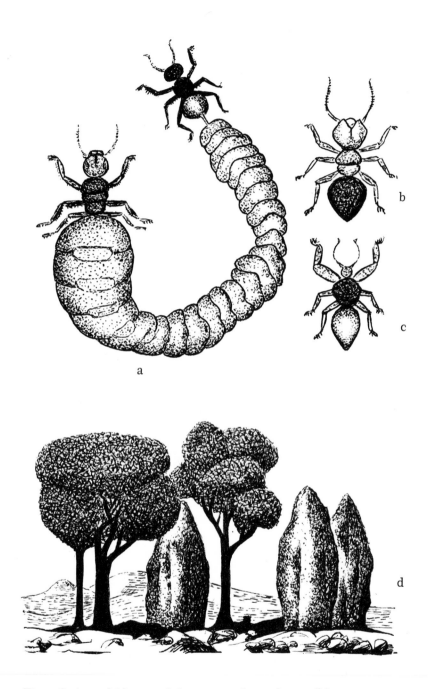

Fig. 18 Antaphid ants: (a) queen and copulator; (b) eater;
(c) builder; (d) ant "forts"

the form of a maze. In the center of the leaf there developed an "alluring" organ which gave off a sweetish odor designed to attract the ants and stimulate their already insatiable appetites.

The antaphids, who like all ants moved along routes generally dictated by environmental conditions, attempted frantically to reach the source of enticement. Running up and down the grooves between the veins they became increasingly neurotic as this apparently simple task came to seem impossible. Every leaf was black with ants thrusting each other aside, climbing over each other, and often killing each other in the grip of a collective frenzy. But what really saved the plants was the fact that the eater ants, in their useless race to gain the middle of the leaf, ate less and less. It thus happened that the builders, and even more the copulators and the queens, who depended on the eaters for all their nourishment, grew weaker little by little and lost the urge for reproduction. In the course of a few decades, mortality began to exceed the birth rate, and in a few centuries the antaphid was extinct. Mastolitz thinks that it was not long after this, and maybe on account of its dramatic victory in the fight for survival, with its competitive drive exhausted by the bitter struggle of evolution, that the plant stood still in time to join that parallel vegetable kingdom in which, with neither growth nor decay, it could maintain its ingenious morphological solutions intact.

In Mali, especially near the villages of Tieplé and Foulan, it is not hard to find fossils of the leaves of *L. labirintiana*. Anyone who has traveled in that region will remember how the roads through the tropical forest of Dang-ma are lined with Dogon boys selling what they call *libi labiliù* to the occasional passersby. Usually these are rough clay copies of a number of impressions taken a few years ago by Tassan and Molheim, and left behind when they went home. The Dogon tribe use the design of *L. labirintiana* for a game, which they call *labi-labi*. They trace the shape, much enlarged, on the sand, using a stick with a rounded point. Then they take turns in hitting balls of beetle dung along the grooves. The objective, of course, is to reach the middle with the least number of shots, although the players know perfectly well that to reach the middle is quite impossible. The game in fact has no winners or losers, but the Dogon play it for hours at a time, without ever quarreling.

THE ARTISIA

For those who have followed the history of the new botany with a certain degree of skepticism, the parallel plants which caused the most perplexity are without any doubt the *Artisia*. This is understandable when we come to think that in the two

PL. XX botanies, normal and parallel, the *Artisia* occupy a very special position, ambiguous because they often seem unbotanical, even nonorganic, and very likely of human origin: this is their dominant feature. When Chabanceau first saw an *Artisia* he is said to have exclaimed: "Ah, enfin une fleur humaine!"

The ambiguous nature of the plant is reflected in its name, which was bestowed on it by the amateur philosopher and botanist Theo van Schamen. It is taken from the gilded inscription which adorns the portal of the Amsterdam Zoological Gardens: "Artis Natura Magistra" (Nature is the teacher of the Arts). Whoever it was who coined this phrase a century ago, when all educated men were still Latinophiles, could scarcely have foreseen that the Dutch would in turn use it to coin a nickname, and call their zoo "Artis." However, it was in homage to this absurdity that van Schamen proposed the name *Artisia* to the Antwerp Conference. He said: "It is not yet clear whether, in its dichotomy of artifice/nature, the plant expresses the influence of nature on art, or that of art on nature."

We know, of course, that it does neither one nor the other, and that apart from its parallelism the *Artisia* belongs totally to nature. But how are we to explain the mystery of those obviously "artistic" forms that in certain specimens we feel must surely be artifacts, copied indeed from the decorative whirligigs of the eighteenth-century baroque?

PL. XX *Artisia*

This phenomenon has been described as "Nature imitating Art," and in the Art News section of the *Aurore de Paris* of January 17, 1973, there was a short article bearing this very title. It ran as follows:

> Anyone who laments the new wave of abstract expressionism which seems to be sweeping through the galleries of Saint-Germain ought to take a look at the small exhibition now set up in the atrium of the Jardin des Plantes. It consists of a recently discovered group of extremely interesting parallel plants. Some specimens can be seen in bronze versions cast directly from the originals by the method known as *plante perdue*, invented by the Veronese foundryman Fausto Bonvicini, and which is simply a new version of the traditional *cire perdue* or lost wax method. Others are displayed with their roots enclosed in plastic cubes of the most crystalline transparency. Others again appear in a segment of their own natural habitat.
>
> Professor Gismonde Pascain, who has been in charge of the parallel section of the Jardin for the last few months, told us that all the plants on show were of exceptional scientific interest. When we asked her which, in her opinion, was the most interesting of all, the young scientist, who was wearing a blue linen dress of decidedly Chinese cut, pointed without hesitation to a group of plants called "*Artisia*," and went on to explain their salient features. To tell the truth, these *Artisia* did not seem to be plants at all, except insofar as they had perfectly real and visible roots. They appeared rather to be worn fragments of baroque chandeliers or of eighteenth-century cornices or frames, picked up for a song, no doubt, at the flea market. Whatever the case may be, they certainly represent a somewhat disconcerting phenomenon which we, who know nothing of the true facts, must attribute to an insane impulse on the part of Nature to imitate Art.

Gismonde Pascain, who has made a thorough study of the *Artisia*, has come to very different conclusions. These are derived from questions which at first sight seem to have more to do with philosophy than biology, and to reflect her connections with thinkers such as Gaston Bachelard and Roland Barthes before she took up the study of biology. She starts by observing that man in his totality is not just *in* nature but part *of* nature. And "totality," for

Gismonde Pascain, includes the important element of his spirituality. "Everything that today is characteristic of man, including his spirituality," she writes, "is the evolutionary result of a series of chance mutations. But in the complex play of infinities these mutations should theoretically be repeatable, just as a royal straight flush at poker is theoretically possible at any moment."

Baldheim's theory that man, rather than being descended from a single source, as is generally held, might have come from a number of sources at various times, is known to be based on these premises. In his book, *Many Adams,*[1] this ingenious American scientist puts forward a great number of interesting theories including the one developed by Gismonde Pascain, which he calls "partial evolution." He thinks that man is the present, transitory result of a series of mutations that in different combinations of order and time might have produced other, different autonomous organisms and living entities. In other words, Baldheim sees man as a mosaic, the elements (*tesserae*) of which might just as well have formed an infinite variety of other images. The theory stems in a sense from post-Darwinian notions of evolution, ideas which, oddly enough, are very ancient in origin. It was in fact Empedocles who stated the first rudimentary principles, using a number of curious images which vaguely recall the physiognomy of several of the lower organisms. "In the beginning," he writes, "there were eyes, and hair, and arms, and fingers. Later on these parts came together, though clumsily at first. Some creatures had eyes in their arms and ears on their hands, while their heads were attached to their legs. Such strange unnatural creatures could by no means survive, and it required an almost infinite number of combinations before the eventual birth of creatures capable of survival."

Baldheim's theories formed the springboard from which Gismonde Pascain made her extraordinary leap of the imagination. Why, she asks, should we rule out the possibility that spirituality might in whole or in part have evolved quite separately from the human shell in which it is housed? Maybe the songs of the birds, and even of the crickets, she says, are simply branches that spring by chance from the great evolutionary trunk that culminates in the music made by man. Nor is it impossible that the ritual dance of the funbirds and of many species of wader are not isolated things, characteristic of a particular species and incapable of further development, but transitory phases in the evolution of dance in general.

Passing from the animal kingdom to that of plants, Gismonde Pascain expresses the opinion that certain flowers, such as *Aracnea ludens,* show some surprising similarities to the decorative head-dresses worn by the people of the Pagunian Islands, which lie to the east of the New Hebrides, proof perhaps that these plants represent a phase in the general artistic evolution which has reached its peak, for the moment, in the artistic products of man. Seen in this light, the analogies between the forms of nature and those which proceed from the creative impulses of humanity take on new meanings. The relationship of art to nature should "reflect the principle that art, as a manifestation of the spirituality of man, does not have an outward and objective relationship to nature but is, like man's body, an integral part of it."

From here it is only a short step to an explanation of the phenomenon of the *Artisia,* which until recently might well have seemed a disturbing coincidence. Gismonde Pascain assures us that the strange and alluring shapes of the *Artisia* are part and parcel of the general evolutionary process of form. They are, so to speak, a lateral development bound to the development of art by having a common matrix.

The *Artisia* on display at the Jardin des Plantes comprise more or less a third of all specimens which have so far come to light. The Botanical Biology Laboratory at Palos Verdes (California) has three splendid specimens *in habitate.* The Laboratorio delle Campora, where Bonvicini made his first casts by the *plante perdue* method, has three plants of the *A. candelabra* variety, complete with roots, as well as the famous group known as *A. magistra,* which was found in the Australian bush by the zoologist Manuel Smithers.

Smithers teaches comparative zoology at Brisbane University, and is also president of the Australian "Save the Kangaroo" Society. He has for some years been leading teams of students into the Australian bush and desert in an attempt to make a count of the few remaining King Kangaroos. It was during one of these expeditions that Smithers saw the now famous group of *A. magistra* in the shade of a eucalyptus tree. Although this extremely competent scientist had never seen an *Artisia,* and indeed was not particularly interested in parallel botany, he at once had an intuition that these were parallel plants, so he warned his students not to touch them. He photographed the group and calculated the exact position. He then sent all his information to his friend Amos Sarno, the most

distinguished Australian botanist of the day, who not only confirmed that the plants were parallel but identified them without the
least shadow of a doubt as *A. magistra*. A few weeks later Sarno arrived with the necessary tools and scientific equipment. He succeeded
in solidifying the soil around the plants and was able to remove
the entire group intact, together with half a square meter of earth.

In May 1974 Sarno went to Europe, and while in Italy he paid a
visit to Professor Vanni at "le Campora." There he much admired
the splendid bronze of a *Solea fortius* which Vanni had modeled in
wax according to the description found in the diary of Amerigo
Mannuccini, a kangaroo hunter who crossed South Australia from
east to west at the beginning of the nineteenth century. Sarno knew
that in Australia the *Solea* had been extinct for some time, and that
all direct evidence of it there had been removed by European col-
lectors. He therefore took advantage of his visit to suggest to Vanni
that they might exchange the *Solea* and the group of *Artisia magistra*. Partly from a sense of guilt, and partly because he could not
resist the temptation to own such an exceptional group of *Artisia*,
Vanni accepted the offer. The plants were dispatched in the autumn
of the same year, but in spite of all the loving care spent on packing,
the group arrived in three pieces. One of the plants (*A. m.* 3),
unluckily the finest of all, was badly damaged and needed very
careful restoration. As the plants were so typically eighteenth-
century in form, Vanni quite rightly decided to entrust the delicate
task to Giovanna Accame, who has the reputation of being Florence's
best restorer of late Renaissance and post-Renaissance works.
The group now looks perfectly intact, and the restorer's hand is
indiscernible.

The *Artisia* of the Laboratorio delle Campora group are fairly
typical of all the rococo specimens yet found. Even the most flourishing specimens, if parallel plants can be said to flourish, are
composed of two kinds of leaves which occur over and over again.
Vanni calls them "involuted" and "devoluted." The involuted leaves
curl in upon themselves in a gesture of introspection which might
be felt to be the prelude to some partial rewinding. The devoluted
leaves, on the other hand, curl in the opposite direction, opening out
in a gesture of offering. All the leaves are one of three sizes, and
there are no intermediate dimensions. This is typical of parallel
plants, which are not subject to the laws of growth. What Vanni
calls *Artisia* are in fact colonies of individual leaves, each one of

which, in spite of "belonging" to a group, has an existence of its own and should be thought of as *an Artisia*. The leaves do not have roots in common, and in cases when they lean upon each other they do so without any functional significance of any kind. We are in fact concerned with an instance of what Gismonde Pascain calls "urban flora," an expression intended to stress the independence and at the same time the solitude of individuals within the group. Other examples of urban flora are the colonies of *Protorbis minor*. On the other hand, the tirils and the woodland tweezers are examples of grouping of the type which Gismonde Pascain calls "collective flora," because they have well-marked social needs and therefore a genuine relationship of interdependence.

A few months after the group of *Artisia* was delivered, Vanni received a long letter from Sarno. Struck by its amazing contents he had it mimeographed and distributed to several friends and colleagues. The letter consists of eleven typewritten pages, the most interesting part of which has to do with the hypothetical origin of the involuted and devoluted curls of the *Artisia*.

> For over two years I have been working with the entomologist Eugene Hopkins, following a line of inquiry into the curves of the *Artisia* leaves, and I must say our researches have led us to some fairly startling conclusions. If I did not mention this during my first visit to Italy, when I so much enjoyed your hospitality, it was because I was still waiting for a definitive reply from Hopkins about a part of his work that he was just finishing up at that time. I hope you will forgive me for my silence on that occasion, and now my dear friend it is a great pleasure to share with you the results of these two years of work. They might even contain the scientific explanation of phenomena previously considered totally mysterious, things I remember that we talked about with such enthusiasm in the magnificent and delightfully Tuscan courtyard of your laboratory.
>
> We have concluded that the leaf-curls are nothing less than the work of a strange insect so far unknown to entomologists, which Hopkins has christened *Artisopteron*, and which might be termed a zoological equivalent of parallel botany—a "parallel insect" in fact. A sensational discovery with unforeseeable consequences!

Artisopteron shows a slight resemblance to certain *Cole-optera*, but at the same time it cannot be classified along with any known genus or species of insect. Like the body of an insect, its body consists of head, thorax and abdomen, and it has six legs. But it completely lacks spiracles, those minute holes which normally form part of the breathing apparatus of insects. The wings, rigid like those of *Coleoptera*, are rudimentary and barely perceptible. Although slightly larger than a common ladybug, the insect is totally invisible to the naked eye.

On January 7, 1973, I decided to carry out a taumascopic examination of our *Artisia*, which then included the *A. magistra* now in your possession, and for this purpose I asked my colleague Hopkins (whose lab is next door to mine) if I could borrow his Somer instrument, the only one of its kind in Australia. On that occasion I found that at the base of the *Artisia* there were a number of small insects, clearly visible by the light of the Tauma-rays. I paid no particular attention, and it was only in the course of a second taumascopic examination three months later that I discovered that if I turned the machine on and off, the insects that could clearly be seen by the Tauma-rays were absolutely invisible without them. I was so amazed that I called Hopkins, and it was then that we started on our research. We are now in a position to give the gist of the first positive results, though we are only too aware that there is still a great deal of work to be done.

Artisopteron lives in the dwarf eucalyptus forests of the bush regions of Knopenland, in eastern central Australia. Attracted, it seems, by the sweetish smell of the trees, which is fairly pronounced in the leaves and secondary roots, *Artisopteron* forms groups of three or four individuals and lives underground among the roots of the eucalyptus as well as at the base of the *Artisia*. It moves extremely slowly, about two steps at a time, usually together with the other members of its group. It has no organs of sight, and as I mentioned earlier it has no real respiratory apparatus. Among the most disconcerting aspects of this creature is the total absence of reproductive organs. In point of fact, throughout the entire two years of our research we have been unable to identify any normal vital processes whatever. At first we were inclined to think that we were dealing with a form of hibernation, but eight seasons have now passed and in the individuals under study we have not observed even the minutest physical change. We now think that we are confronted with a physical condition which cannot be defined either as life or as death. In this respect *Artisopteron* is very similar to certain parallel plants, such as *Artisia,* which are motionless in time.

But the feature that has struck us most is a minute stinger at the bottom of the abdomen. In the light of the Tauma-rays this shows up with intense brightness, of a color that varies with the individual from cinnabar red to emerald green. We thought at first that this was a sexual differentiation, but further experiments revealed that there is a direct relationship between the color of the stinger and the shape of the leaf on which *Artisopteron* lives. In brief, we found that the insects with the red stingers live on *Artisia* with devoluted leaves, while those with green stingers live on the plants that have involuted leaves. The simplest hypothesis was naturally that the insect somehow punctured the leaves, thus causing the directional development of the curls, but in the course of two years of intense study we have been unable to discern any direct causal relationship beyond the simple fact of their presence on the leaves. We know from our experience in the field of parallel botany how powerful the effect of this presence could be, and we therefore came to think that the curl of the leaves was determined by the mere existence of either "red" or "green" *Artisopteron* on the plants. This naturally does not exclude the possibility that between insects and plants there might be a mutual attraction, a simple *a posteriori* selective relationship.

This is the point we have reached at the present moment, my dear Vanni, but we intend to continue working on the specimens of *Artisia* we have in our possession, which luckily are quite a few, as well as those still hidden in their original habitat in the shade of the dwarf eucalyptus.

We can hardly be surprised that Vanni was shaken to the core by this letter. What is really astonishing is that no one before Sarno had ever seriously considered the possibility of a parallel fauna, even in the case of insects, which have such a close symbiotic relationship with plants. It is too early to make any forecasts, but we cannot help thinking that the latest news from the Antipodes justifies some expectations heavily loaded with suspense.

The phenomenon of the curling of the *Artisia* leaves has other interesting facets, the most curious of which is without doubt the PL. XXI Kaori tattoos. The Kaoris must be considered the first real settlers or colonizers of the Australian continent. They landed there after the Chinese, between the thirteenth and sixteenth centuries, but unlike the navigators of the coasts of Asia they pushed on into the interior and established themselves permanently there. They came from the islands of Polynesia and brought with them the tradition of tattooing. In their novel environmental conditions, confronted with natural forms that were new to them, this tradition of theirs underwent profound modification. The highly elaborate tattooing of the Kaoris is in effect a marriage between extremely ancient Polynesian forms and Australian themes of more recent origin. That the Kaoris knew the *Artisia* and attributed magical powers to the plant can clearly be deduced from certain details of the tattoos and from the paintings on bark which have been meticulously documented by the Department of Anthropology of the University of Brisbane.

The general purpose of this tattooing is to integrate an individual within a social group. But it also signifies the symbolic conquest of the things represented. In the case of the *Artisia* it is now almost certain that to the Kaoris the plants represented the annulment of time, and hence eternal life. That centuries ago a primitive people was able, if only intuitively and with the attribution of supernatural meanings, to discern a phenomenon that is only now being timidly explored by Western science is indeed a most extraordinary fact.

In a letter to the Anthropological Society of Australia, Professor Anthony Campbell explained the magic significance that the *Artisia* have for the Kaori people and described the tattooing ceremony which takes place—and not by chance—in a hut built of eucalyptus boughs. The rite is presided over by the shaman of the tribe, but the act itself is performed by the *astòk*, a kind of itinerant artist possessed of magic powers and a special skill in tattooing. At one time there were many of these *astòk* traveling about in the Australian bush, but today it is a dying profession, kept fitfully alive by subsidies from the Department of Kaori Affairs.

The ceremony takes place once a year and involves the whole tribe. All the young who have reached the age of twelve in the course of the year are tattooed, regardless of sex. Only the face is tattooed at this stage, the rest of the body being done later.

PL. XXI Kaori tattoos

The *astòk* begins his work by taking a charred eucalyptus twig and drawing two *Artisia* leaves, one on each cheek. Around these he then adds the intricate designs which follow the form of the face and accentuate its individual character. Most of the lines are abstract, but they may also be symbolic. Sometimes two tiny *Artisia*, one involuted and the other devoluted, are represented on the sides of the nose. While the *astòk* is at work the elders of the tribe all squat round the circular wall of the hut, which is about eight meters in diameter and festooned for the occasion with thousands of bright-colored threads of wool hanging from the vaulted roof. The men sing a monotonous rhythmic chant, which is in fact the invocation "Atnàs-poka-nama poi" (Great mother of the long night), while outside groups of young people hold hands and dance around the hut to the same rhythm. Every now and then they brandish eucalyptus boughs and shout "Pokà." When the drawing on the face is finished, the old men leave the hut and the *astòk* begins to execute the tattoo itself. This is a painful process, and as the designs are so extremely intricate it can last for a whole day. At one time the skin was punctured with the thorns of *Solicarnia pendulifloris*, but in the early days of British rule the *astòk* began to use ordinary sewing needles, manufactured in Birmingham and obtained from English travelers in exchange for the kangaroo skins then much in fashion in Europe.

As I mentioned above, *Artisia* also appear occasionally in bark paintings, which have recently acquired some fame with the growing interest in primitive art. One such painting on exhibit in Paris at the Musée de l'Homme clearly represents a large *A. major*, devoluted in form, between the two figures of a kangaroo and a hunter.

In a short essay recently published in the *Annales* of the Musée de l'Homme, Gismonde Pascain points out that to the Kaoris the two forms (involuted and devoluted) represent the inner and outer parts of man, that is, body and soul. When they are represented together, as in the facial tattoos, they stand for this dichotomy. In the paintings, however, there is nearly always only one *Artisia*. In the particular case of the painting in the Musée de l'Homme, she says, the form is devoluted, expressing more concern for the body than for the soul. Involuted forms occur seldom in Kaori iconography, according to this leading French biologist, a fact which bears witness to the sense of realism and excellent mental health of the natives of Australia.

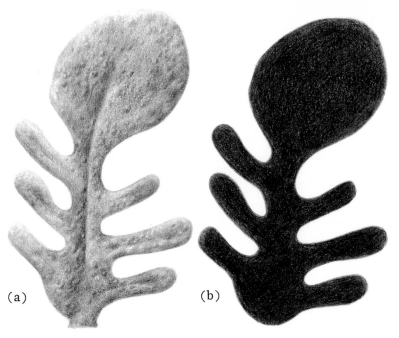

(a)

(b)

Fig. 19 (a) *Artisia Arpii* and (b) a collage by Jean Arp

(a)

(b)

Fig. 20 (a) *Artisia Calderii* and (b) a pendant
by Alexander Calder

In dealing with the *Artisia* we have often had occasion to mention their typically eighteenth-century forms. It is perhaps only to be expected that a period so rich in all kinds of representation of flowers should furnish us with easy comparisons. But we ought to bear in mind that a number of specimens were known before the eighteenth century, even if their parallel nature was not then suspected, and also that many *Artisia* reflect the styles of other epochs. We need only mention the so-called "Carolingian" *Artisia*, which bears so great a resemblance to the magnificent bronze plaques of the doors of San Zeno at Verona; this plant is now in the little museum at Casteldardo, where it was found over a century ago at the foot of the age-old eucalyptus whose massive dignity still dominates the tiny public gardens of this pretty little town. And to turn to more recent art, we should not forget *Artisia Arpii,* which owes its name to the amazing similarity of shape between it and certain collages and pieces of sculpture by the dada artist Jean Arp (Fig. 19), and *Artesia Calderii,* whose motifs irresistibly recall those in the work of the late American sculptor Alexander Calder (Fig. 20). Indeed Jean Alembert, art critic for *Les Jours,* has gone so far as to write that the day will come when a single display of parallel botany will embrace the whole complex panorama of Western art from its beginnings down to our own days.

THE GERMINANTS

Rather than being plants, the germinants are a combination of heterogeneous elements of which the really parallel part is perhaps only minor. They have no proper botanical gestalt, and lack that overall plantness that is one of the most obvious features of the other parallel plants.

The name *germinant* was coined by Jacques Inselheim of Strasbourg University. During a trip to Italy he was considerably struck by a number of plants which he saw at the Instituto Venturi in Cadriano, near Bologna, and as soon as he returned to France he wrote an article for the *Gazette de Strasbourg* in which he described his encounter with this unusual flora. As they had only just been discovered at that time they had not been given a name, and thus in a moment of weakness and enthusiasm, certainly questionable from the strictly scientific point of view, he called them "germinants." How are we to interpret this name? Is the word transitive or intransitive? Are we concerned with "that which germinates" or "that which is germinated"? When asked about it at last year's Baden Baden Conference, Inselheim explained that the ambiguity of the name was the result of an absolutely intentional choice, and that he was only too glad to take full responsibility for it. If it is true, he said, that the term *germinant* refers to something which germinates buds, it could equally well refer to buds which germinate. He was struck by the similarity between a verb that, absurdly enough, can be transitive or intransitive, and a plant that appears to be generated as an organ by another plant, but which in reality is entirely separate and complete in itself. "The germinant," he said, "is beyond doubt the most ambiguous of plants. And it is only right and proper that it should have the most ambiguous of names."

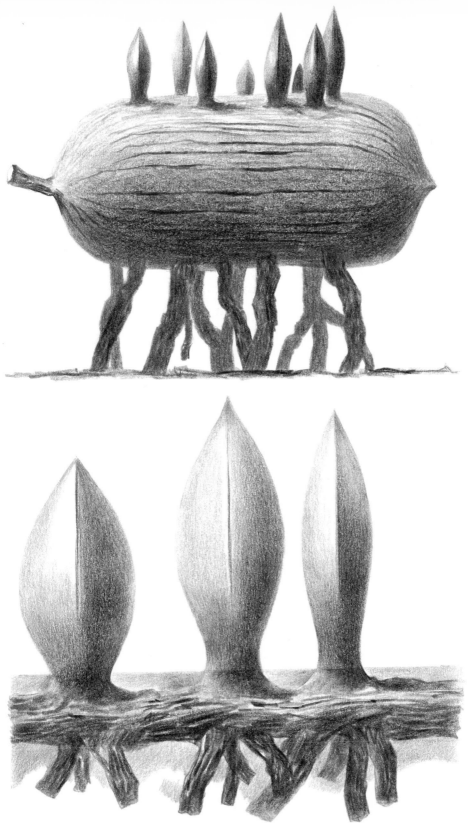

PL. XXII The Cadriano germinants

Inselheim saw two of these plants at Cadriano. The first resem- PL. XXII bles a large squash standing raised on about twenty scraggly and irregular roots of the type known as *ambulans*. From the rough skin of the *cucumbra* (generating body) there sprout a dozen arrogant buds, which are shiny and perfect: the germinants. In the other specimen the buds (also a group of twelve) spring from an aquatic rhizome about forty centimeters long which has been successfully enclosed in a block of polyephymerol.

Following his visit to Cadriano, Inselheim bought a single germinant from an amateur botanist in Bologna. This sprouts from what appears to be a bit of volcanic rock about the size of a clenched fist. So far it has not been identified. In any case, Inselheim presented it to his alma mater, the University of Padua, in memory of his old teacher Professor Alfonso delle Serre.

The two groups of germinants at Cadriano are almost identical, even if the different elements from which they appear to grow display them in very different contexts. The better known of them, which scholars refer to as the "Cucumbra" germinants, has aroused endless problems and disputes wherever it has been the object of study. The special report prepared by the Faculty of Botany in Bologna is in fact in clean contrast to the opinions held at the Cadriano Center. The latter are based on a premise that seems to us scientifically correct, that in parallel botany there are no organic connections between the various parts of a plant. When these parts display an apparently arbitrary relationship, as in the case of the "cucumbra" germinants, then the only reasonable method of study is to take the parts separately, without prejudgments, and explain their coexistence as best we can.

Following this structuralist procedure the Institute has arrived at these conclusions. The twelve germinants are certainly and unequivocally parallel. Irrefutable proofs are furnished by the continuity of their internal substance, their morphological inalterability, their tendency to turn to dust on contact with foreign bodies, and the strange behavior of their image when recorded on film.

The mother *cucumbra*, as Inselheim calls it, does not on the other hand seem to have the qualities that would enable us to call it parallel. The fact that the plant was discovered in the neighborhood of Ferrara, near the Certosa di Pomona, in a thick hedge surrounding a field full of summer squash, justifies us in entertaining reasonable doubts. Furthermore, experiments using minipolarization

have shown that the cucumbra reacts to external agents exactly like
any normal fruit. In theory it would allow itself to be cut into slices
and at high temperatures its substance would undergo considerable
alteration. It was these considerations in the first place that induced
Professor Giancarlo Venturi, the founding father of the institute, to
judge the mother *cucumbra* to be an anomaly belonging to normal
botany.

For the scientists of the University of Bologna, however, the
germinants were originally real buds that sprang from the *cucumbra* and have now entered a condition of parallel stasis. The fact
that the *cucumbra* looks like a zucchini and the circumstances of its
discovery are considered to be pure coincidence. They point out that
probes into the interior of the fruit carried out by the Anten-Abrams
method have not revealed the presence of seeds or even the least
variation in the density of the material. What appears to us as the
skin from which the germinants emerge by breaking violently
through it, is nothing other than the external limit of the interior
substance. The germinants are firmly attached to it, so much so as
to seem of the very same substance. The surface irregularities such
as protuberances and longitudinal scratches are, according to the
Bologna botanical team, of paramimetic character.

Professor Mario Federici, who drew up the report, taking into
account all combinations in which the germinants figure as a single
and distinct parallel entity, tends to minimize the importance of the
buds in favor of the matrix, and he speaks of the "germinating
cucumbra" and the "germinating rhizome." He describes even the
"ambulant" roots of the *cucumbra* as a parallel phenomenon, although he recognizes that certain features recall the petrified plants
in the Chuhihu Valley.

Venturi, on the other hand, is of the opinion that the roots belong
to normal botany and in the case of the *cucumbra* are no more than
a fortuitous circumstance. He says that originally the *cucumbra* lay
on the ground like an ordinary zucchini. Underground roots were
attracted by the damp which was to be found in its shadow, and
converged upon it. Then by a slow process of penetrating antiastasis
they ended by raising the *cucumbra,* detaching themselves from
their original root system and transforming themselves into the
ambulant type by a series of later mutations.

The two theories are equally divergent as regards the aquatic
germinants, and although there are two elements rather than three,

the reasons for attributing them to one botany or the other remain the same. For Venturi the underwater rhizome is just a rhizome *capirens* in the process of parallelization, while for Federici it is part of a single parallel entity. Where the two scientists agree entirely is over the attribution and hymothesic description of the buds. Both admire the high level of ambiguity of the plants, and wonder whether it is a case of a sudden stoppage of development at the moment of parallelization or of a precise gestaltic intention. This question was discussed at great length at the Baden Baden Conference, and the majority of the scientists present favored the second view. The buds, rather like the seeds of *Giraluna*, would appear to represent what in human terms would be called an "idea." They are the programmed and definitive form of a meaning, a "design" by nature, we may say. This coexistence of a content, a narrative, with the simple phenomenon of self-presentation, is possible only in parallel botany. The resulting ambiguity is due to the apparent incompatibility of time, without which an idea cannot exist, with non-time, which is the *conditio sine qua non* of the plants found on the other side of the hedge. The germinants, with their seemingly vital impulse which presupposes a history and suggests a future, are pointed aggressively at the sun, like missiles programmed to strike at and explode the last (or the first) mystery of living matter. But their inert matterlessness, their immobility outside of time and their being only illusorily set in space—these qualities exclude them from having any part in the growth and development of the biosphere. Theirs is an existentiality of dreams, in which form and meaning are a single materialized fiction, suspended between the light of our perception and the darkness of their own being.

Inselheim holds the view that the germinants are an Italian plant, and he supports this theory with a great deal of paleontological, geological, meteorological, and toponomical data. It is true that the only germinants yet discovered have been found in Italy. After the three specimens already mentioned, other plants have been seen or obtained from the Gargano, from Castellina in Chianti and from Rocca di Faggio. The Natural History Museum of Verona has two specimens of the *cucumbra* type recently acquired from a small farmer in Caselle. There is every reason to believe that the germinants are not only a specifically Italian plant, but that of all parallel plants they might well be the most numerous and easily accessible. But unfortunately Italy is the only country which still has no laws to govern and protect parallel plants, and no provisions whatever to encourage research. As they are not plants in the usual sense and as it is difficult to define their nature and substance in legally acceptable terms, their continued existence is seriously threatened by the vandalism of weekend vacationers, as well as by the selfishness and ignorance of amateurs and speculators.

In the meanwhile, Colonel Di Bonino of the Forestry Police refuses to accept responsibility for things that do not form a part of the vegetable kingdom. Under-Secretary De Francisci, who is responsible for ecology within the Instituto per lo Sviluppo Economico (I.S.E.), was appealed to by the University of Bologna, but replied in extremely vague terms and did his best to make the problem appear ridiculous. Senator Giuseppe Montaldin, president of the Committee for the Defense of the Products of the Soil, recognizes the scientific importance of the germinants but denies that they can be called products of the soil, while Giovanni Amarà of the National Scientific Research Institute (I.N.R.S.), in a memo to Minister Fratelli which on the whole was sympathetic and reasonable, lists his reasons for being unable to intervene, including shortage of funds, lack of qualified personnel, and, above all, the troubles which would accrue to the institute if it concerned itself with a problem which could not be explained clearly and simply to the politicians who control its activities.

THE STRANGLERS

B etween the two groups of plants which comprise parallel botany as we know it today there is a mysterious no-man's-land in which vegetable organisms, now extinct, once lived out an anomalous existence.

The plants are exceptional in form, behavior, and orthogenesis, and cannot be placed anywhere in the existing classification of parallel botany. They have for this reason been the object of special study by botanists, paleobotanists, psychologists, and even poets.

One genus in particular, the so-called "stranglers," exemplifies PL. xxi all the features of that small group of plants which scientists have christened the "phab" group (from al*pha b*eta). The existence of the group was discovered a few years ago by a team of paleontologists led by Professor Ahmed Primshama of the University of Baroda. While working in the hills near the Jain temple of Mount Abu they came across some fossils of hitherto unknown plants. It seems that they were examples of a type of tiril, about thirty centimeters high, that must have been quite common to the north of the Indian subcontinent for some millennia at the end of the Orthoplantain era. Fossils that came to light in 1971 in the Shetford coal seam, and were assigned to a much more recent period, show surprising morphological analogies with the Indian specimens. Primshama himself agrees with the English paleobotanists Smithen and Mc-Cook that they are "strangler tirils."

These scientists have put forward some original and convincing hypotheses regarding the life of these plants, which must have been endowed with abnormal vital urges. Smithen and McCook hold the view that in order to limit their distribution nature provided them with a curious mechanism of ecological control, without which in the course of two or three million years they would have covered whole continents at the expense of all other forms of life. According to these experts this consisted in a quite exceptional self-destructive aggressiveness developed during the flaringean phase of growth, which the tirils expressed by slowly and gradually winding themselves round nearby plants, even those of their own species. The extinction of one species by reciprocal strangulation—the process which botanists call *eronecria*—must have taken place in the course of a few millennia, but before total destruction some specimens must, by mutation, have generated a new species also genetically equipped with the suicidal instinct. And so on and so forth. The last survivor of the long series of stranglers was probably *Tirillus maculatus*, which was far less aggressive than its ancestors must have been. Even today this tiril, destined to outlive all the other stranglers, covers vast areas of the Alaskan tundra, where it is a favorite food of the herds of dwarf caribou that sweep down into the peninsula every spring.

With regard to the stranglers, Von Harne recently published a sensational article in the *Archives of Parabotany*. His theory is that in the history of parallel botany there have been numerous other plants which have disappeared from the face of the earth, only to reappear at some distant time and place, slightly modified in form and behavior. He formulates the theory of a vegetal metempsychosis due to which the processes of life and death remain suspended in the impressions of capillary roots "burned" by time into petrified veins, clays, and crystals; and that these transmit generating energies by means of an imperceptible, age-old osmosis. The genes, freed at last from their long subterranean slumber, pass on the ancient existential programs to new plants.

Von Harne pays particular attention to the stranglers, tracing a long history "which like a distant archipelago appears to float in time." The "soul" of these plants seems to be the agent responsible for the infinitely slow violence which is their main feature, and according to Von Harne this survives in species such as the common ivy, thus revealing unsuspected links between the two botanies.

PL. XXIII Strangler tirils

THE GIRALUNA

This elusive and capricious plant is the Dream Queen of parallel botany. Hydendorp, quite rightly, does not hesitate to define it as the "most parallel of plants, most plantlike of the parallels," and in so doing he stresses not so much its physiognomic oddnesses as the disconcerting *normality* of its shape. "If we were in the jungle," he writes, "and we found one blocking our way, we would not for an instant hesitate to hack it down with our machetes."[1]

But it will not be our good fortune to encounter it. If in reconstructions the *Giraluna* displays considerable plantness of form and PL. xxiv an exact and convincing solidity, in its natural environment it can be perceived only as a nebulous interplay of glimmerings and empty spaces which alternate in the darkness and vaguely suggest where its outlines might be. Its nocturnal presence, in fact, is manifested almost entirely in terms of the equivocal O'-factor of the moonbeams, which was discovered and measured a few years ago by Dennis Dobkin of the Point Paradise Observatory. This factor changes the light-shade ratio which normally defines volumes into a subtle interplay of lucencies and opacities, so that our perceptions, our basic sensorial habits conditioned by thousands of years of daytime life in the "solar key," would need complete readjustment and indeed reversal in order to come to terms with it. Daylight isolates objects, bestowing a noisy meaning on all the odds and ends in the world. But night takes everything away except the very soul of things: a black light, a transparent darkness, a secret we cannot grasp.

During the long night of the Erocene era man caught a glimpse of the *Giraluna* rising mysteriously in its barren landscape. Presolar man imagined himself the child of the Moon. In her lap he had known the comfort of the idle, silent torpor of the night, and by her light he had seen silver pearls lie weightlessly upon the coronas of the first great flowers. But he left us only a few enigmatic signs of all this: the Feisenburg cave, the petrified bones in the Ahmenstadt tumulus, the Boergen Cup. Paradoxically enough, all that we do in fact know of his presence in that landscape comes to us from our study of his nocturnal vegetation.

Around the middle of the Erocene era, when the flowers of night were fading away in the light of a new dawn, man saw that outlines and colors were slowly hardening. Thus he discovered the stone-hard world of day, and learned to be the child of both Sun and Moon, of Amnes and Rā, of Disana and Karak, of Nemsa and Taor. The "crawling stones" of Yorkshire, the stele of Tapur, the graffiti of Klagenstadt, these have preserved for us the nearly obliterated images of the two divinities who from the center of their temples drew the design of the universe.

But the Sun was not long in attaining absolute power over everything in the world. "O Rā, O Amno Rā our benefactor, glowing and flaming! Gods and men bow down before you, for you are their creator and their only Lord." Such was the prayer of Amresh, High Priest of Egypt. And a new vegetation, outspoken and exuberant, appeared on the earth, and made the bright leaves dance in the morning breeze. Night soon became no more than a dark corridor joining one day to another, a place of visions and memories, a storehouse of words and images. It became a secret refuge where the vanished flowers could once more flaunt their coronas to the Moon. And thousands of years later the black flowers of that distant night—*Giraluna, Lunaspora, Solea argentea*—were born from seeds hidden deep in a soil rich with legends and stories.

If our knowledge of the *Giraluna* is today reasonably complete and detailed this is due to the industry and scholarship of Professor Johannes Hydendorp of the University of Honingen, who has collected and collated all known facts and kept his records abreast of the latest developments. Our historical and geographical information comes from the most varied sources: legends and folk tales handed down from generation to generation, accounts given by explorers, anthropologists, and paleontologists, and of course the more recent testimony of botanists such as Heinz Hornemann and Pierre Maessens.

PL. XXIV *Giraluna* (closeup of *avvulta* at right)

The iconography of the plant includes the Solingen graffiti, the polychrome bas-reliefs from the Karno tombs, and the clumsy sculptures made by the natives of Uranda. But of fundamental importance is the recent discovery of the great *Giraluna* of Somma-campagna near Verona, which is fashioned in such a meticulous manner as to carry the utmost conviction.[2] Also important is the Lady Isobel Middleton group, which we will discuss below at some length.

By interpolating descriptions and representations, Hydendorp was able to draw up a really detailed morphology of the plant, which now requires only the confirmation of direct observation. Although his own reconstruction was made a year before the discovery of the *Giraluna* at Sommacampagna, it is identical in every detail with that colossal bronze—fair proof of the accuracy of both.

Hydendorp lists three types of *Giraluna:* the common variety, *G. vulgaris,* typified by the Sommacampagna bronze and which has all the basic characteristics of the plant; *G. gigas,* a native of the mountains of Tampala in India, and which can reach heights of three meters or more; and *G. minor,* found in the undergrowth, a plant rather similar to the fungoid *Protorbis minor* and occasionally mistaken for it. The three varieties all greatly differ in size and substance, but they share the pendulous roots and the circular corona with its spherical seeds that shine with a strange metallic light.

Although the *Giraluna* is not a social plant, there are groups of the giant variety consisting of three or more individuals which, as with normal plants, appear to be connected by means of a common root system. *G. minor* is sometimes found in even larger groups (Hydendorp mentions up to forty plants in one square meter) which would lead us to think, quite erroneously, in terms of a rudimentary social structure such as that of the tirils or the woodland tweezers. But in fact it is no more than chance association without any suggestion of interdependence and due merely to favorable environmental conditions.

In his description of the *Giraluna,* Hydendorp distinguishes two main parts, the "trunk" or "column," and the "corona." The trunk corresponds more or less to the stem or stalk of more delicate plants, but deserves the term applied to it on account of its unusually massive bulk in relation to the plant as a whole. A fully grown *Giraluna* measuring a meter and a half in height can be as much as forty centimeters in diameter at the base. The trunk of the *Giraluna* is composed of the column proper (Lat. *columna*) and a system of aerial roots, the pendulants, which are collectively known as the *avvulta*.

As with most of the parallel plants of the Beta group, the column rests on the earth without in any way being attached to it. In spite of this, the plant stands so steadily that in the absence of the appropriate technical apparatus it is impossible to knock it over or to cut it down. Here and there on the surface of the column there are fragmentary remains of a corklike bark. This is an extreme case of "parabotanization," a type of camouflage aimed at concealing the parallel nature of the plant. It has been the subject of a special study by Hydendorp, who attributes the phenomenon to psychobotanical deviations which probably resulted from a malfunctioning of the genetic memory. Mutilations of the mnemonic system or, in plain terms, cases of forgetfulness, would seem to be the cause of anomalies for which we can find no other explanation. Hydendorp calls our attention to the theories of Hermann Hoem, according to whom false messages can be delivered, while mistaken decodification and even genuine forms of plant neurosis can take place in the evolutionary process at cellular level. He cites some well-known examples of plants which have undergone surprising morphological changes by means of rapid mutations caused by frustrations, inhibitions and obsessive envy. Here we might mention the clear impressions of roses to be found on the leaves of ivy of the variety *Rosa alienis*, the shapes of flowers and leaves that appear from time to time on the cracked bark of *Pinus adelphis*, and finally the numerous "fruit-bearing leaves" which result from a long series of laboratory experiments carried out by Hoem himself.

The upper part of the column of the *Giraluna* is much narrower than the base, and is curved in such a way as to hold the corona at the proper angle, usually forty degrees. At the very base it swells as if to give the plant more stability, and in this it is aided by the lowest pendulants, which often extend onto the ground for ten centimeters or so. In fact, of course, this swelling at the base is also part of the camouflage game of the *Giraluna*, which pretends to require physical and gravitational stability while in fact its remarkable "solidity" proceeds from the very nature of its matterlessness.

Apart from the fragments of pseudobark the column has the smooth and slightly viscous surface characteristic of all parallel

PL. XXV The *eblùk* procession (Sumerian bas-relief)

plants. Sometimes, like certain varieties of tiril, it is covered with a very thin layer of faintly scented wax, known as emyphyllene, which is reminiscent of the copaiba resins of certain normal plants. The men of the post-Erocene era thought that this substance had aphrodisiac properties. Maessens is of the opinion that the famous *eblùk* of the Sumerians, used by the sculptor priests to anoint the block of basalt from which they were to draw forth the image of the emperor, was nothing other than the emyphyllene of the *Giraluna*. In those days the plant grew in some abundance in the shade of those vast rocks which the Sumerians took to be pieces of the moon that had fallen into the desert. On nights when the moon was full long lines of warriors, led by the priests, would scour the desert in PL. xxv search of the precious balsam, which was heated and poured into a small golden casket. At the climax of a complicated nocturnal ceremony, while the High Priest delivered rhythmic blows of the hammer upon the holy chisel, the contents of the casket were poured over the shapeless stone which was to yield the portrait of the defunct sovereign. The balsam was supposed to give the dead emperor such virility that in the other world he was able to re-create his entire people. According to Maessens the stele from the Karno tomb, which shows a pair of mythical animals, half lion and half bull, supported by a stylized *Giraluna* (Fig. 21), also exalts the aphrodisiac properties of the *eblùk*.

All around the column hang the tubular roots known as pendulants. These are scarcely visible at the top of the plant, but they become longer and longer toward the base where, as mentioned above, they often protrude for some little way over the ground and provide totally fictional support. The pendulants are smooth, with perfectly rounded ends, and the common variety possesses some hundreds of them. They are similar to the pendulants of the *Sigurya*, but while these are thin, irregular, and likely to appear at any height on the stem, the pendulants of the *Giraluna* are very regular in form and distribution. When several layers are superimposed they form the *avvulta*, which resembles a kind of skirt.

Hydendorp thinks that the way the pendulants are arranged suggests that the *Giraluna* was once entirely independent of the earth. "A plant that has roots falling toward the earth," he writes, "has certainly not grown upwards from beneath the surface, as occurs with normal plants. It is probable that before the beginning of the Metrocene era (the "age of measurement"), when living organisms were relatively few and without dimensions, the *Gira-luna* was an aerial plant."

Hydendorp goes on to remind us that the problem of survival does not apply to parallel botany, and that therefore these plants have no necessity whatever for maintaining contact with the earth. "It is not impossible," he writes, "in earlier times, when the ratio between the gravity of the earth and that of the moon was not the

Fig. 21 The *Giraluna* of Karno

same as it is today, that the earth rejected certain organisms which at that critical moment in the formation of the biosphere might have seemed of no use to the emergence of the new ecological balance. It is possible that great numbers of parallel plants, like their ancestors the *Lepelara,* circled the earth while awaiting mutations that would permit the earth's gravitational force to prevail. The *Giraluna* would then have descended to our planet and stretched out its pendulants like tentacles to help it make a good landing and aid it, at least in the early stages, to keep its balance."

Maessens pushes Hydendorp's theory to the point of paradox, and even speculates on the possible selenogenesis of the plant. For him the *Giraluna* of Sumerkan must have fallen from the moon at the same time as the huge boulders in the Ahem-Bu Desert. There was an absurd but memorable bickering match between the two scientists on this very subject at the 1968 Antwerp Conference. This should scarcely be wondered at, in view of the fact that parallel botany as a whole, and the Beta group in particular, disobeys the laws of nature and therefore encourages speculations that sometimes threaten to compete with the boldest imaginings of science fiction.

If the pendulants reach toward the earth, the corona of the *Giraluna* is definitely turned toward the sky. It is a large circular dish full of metallic spheres that are commonly called seeds, though obviously they are nothing of the kind. The corona of the common *Giraluna* is forty centimeters in diameter. Unlike the varieties *G. gigas* and *G. minor* it has a few triangular petals irregularly placed around its rim. Hydendorp lists this as one of the paramimetic features of the plant, though he also thinks that this might represent a distant "memory," a rudimentary vestige of a previous form of corona.

The seeds of the *Giraluna,* more correctly called "spherostills," differ from one variety to another, but within each variety they are of fixed and invariable size, quite independently of the size of the individual plant. Those of the common *Giraluna* measure four millimeters in diameter, those of *G. minor,* two millimeters, while the spherostills of the *G. gigas* measure as much as twenty-four millimeters in diameter. While the plant is matterless and therefore has no specific gravity, in the case of the spherostills one can speak in terms of an actual mass. This fact was used by Hydendorp to combat Maessens's theory of lunar genesis. "In the fall from one gravitational field to another," he said, "the spherostills, which are not connected to the corona in any way, would have gotten lost in space." In the reply he made during the famous debate mentioned above, Maessens drew attention to the fact that in certain well-known instances the corona is in fact missing a number of seeds, and he declared that the mysterious atmoliths that damaged the lunar screen of the American Macron II might well have been *Giraluna* spherostills which had remained in orbit.

If the fragments of "bark" on the *Giraluna,* and the pendulants which compose its *avvulta,* can be explained by the theory of para-botanization, the so-called seeds, with their geometrical perfection and brilliant shiny surface, appear to elude any rational explanation whatever. But the fact that they are there on the corona leads us to suspect some hidden meaning beyond the scope of any acceptable hypothesis. In spite of their obvious uselessness and their oddly mechanical appearance, we are bound to admit that the spherostills must be in some measure "organic by association," so that in the context of the "plant mother" they are instantly interpreted as seeds. The extent to which the context of the plant is decisive as a semei-otic element has been shown by Anselmo Geremia of the Natural History Museum of Vicenza, who for some years now has been working on the strange phenomenon of the "seeds" of parallel plants.

We here give the results of a test carried out by Geremia to ascertain the recognizability of the spherostills of the great bronze of Sommacampagna: Removed from the context of the *Giraluna* and displayed instead on the workbench of a machine shop, the spherostills were identified as ball bearings by ninety-eight percent of the people taking part in the test. When the spherostills were scattered here and there on the soil in a garden the outcome was more or less the same. But when they were seen in their proper place, on the corona of the flower, the results were completely reversed: ninety-four percent of those interviewed took the spheres to be the seeds of the plant, even if in their comments ball bearings were often used by way of comparison.

But what could possibly be the function of abstract seeds, quite clearly incapable of generating new plants? What relationship could there be between the spherostills and the plant mother with which they have not the least trace of a physical bond? How can we explain the fact that the most "botanized" of the parallel plants has seeds which can easily be mistaken for a product of advanced industrial technology? Geremia writes:

> It is clear that in their distant past the spherostills cannot have had any true biological functions. To demand that plants which enjoy timelessness should take the trouble to ensure the survival of the individual, that they should possess mechanisms to bring about the preservation of the species, is a patent and paramount absurdity. Having also rejected the idea of an extreme form of paramimesis, on account of the mechanical appearance of the seeds, we are forced to give serious consideration to the hypothesis of a purely symbolic function similar to that of the brightly colored tail of the *mitlachec*, a small bird of prey found on the Carador coast of Bolivia, which Hayman Harris defines as a "temptation to the dream." As with the germinants, the highly artificial perfection of the spherostills, in dramatic contrast to the plantness of the rest of the *Giraluna*, brings to mind meanings involving a certain amount of ideality; but ideality implies a hypothetical future, and the future is conceivable only in terms of constant temporal motion. It therefore becomes natural to wonder if the spherostills might not represent a symbolic bridge between the two types of time, the mobile time of normal botany and the motionless time of parallel botany.[3]

The poetic question put by Geremia thus translates the mystery of the physical appearance of the spherostills into new conceptual terms. But what on earth could be the message that the spherostills are carrying from one time to another, from one flora to another? Geremia does not tell us. He is wise enough to think that the answer to such a question must completely elude our logic. Even to ask the question is rather like trying to jump over one's own shadow.

From the steppes of Yaghuria, from the Novaho Desert, from the tropical forests of Central Africa, from all those distant places of the earth where man has seen the seeds of the *Giraluna* glinting in the night, travelers have brought back stories of enchantment and magic inspired by the bewitching mysteries of the plant. Among the most curious of these is a story told by Vladimir Oncharov, a little-known Russian writer,[4] who in 1868 published a volume of children's stories which has never been translated. One of these stories is called "The Flower with the Golden Seeds." Oncharov himself admitted that the story was based on a very old Yaghurian legend, and it stands as proof that even in the dreary landscape of the Siberian steppes the *Giraluna* must once have offered its glittering spherostills to the moon.

Although Oncharov was the first to use the name *"Giraluna,"* and although there is no Russian who has not read or been told the story as a child, no one ever dared to advance the idea that the precious flower might really have existed: an example of how folk tales, so rich in historical, geographical, and scientific information, still remain largely and inexcusably unexplored.

Here is the text of the Russian story in a translation by Giselle Barnes, to whom we are also grateful for bringing it to our notice.

The Flower with the Golden Seeds

Ivan Antonovitch was a poor peasant who lived with his wife Katyusha in a wooden cabin not far from the village of Blansk. What little he earned came from the milk of his ten goats and from selling the seeds of the sunflowers which late in the summer droop their heavy heads as if tired of waiting for the sun, searching the gray earth instead for a reason to live.

One cold and misty September day Ivan Antonovitch decided to reap his sunflowers, which now were ready to be harvested. He sharpened his *amovar* and off he went to the field. He had already cut a few bundles when suddenly he stopped in his tracks: he stood and stared in amazement at the plant he had just been on the point of cutting down. It was the only one in the whole field that still had its face upturned to the sky. It had black petals and seeds that looked like tiny nuggets of gold. Ivan Antonovitch rubbed his eyes and pinched himself hard on the nose. Persuaded at last that he was not dreaming, he carefully gathered the heavy seeds and dropped them into the huge pocket of his smock. He continued to harvest his sunflowers, but every so often he took a seed from his pocket to assure himself that it really was gold. Once he nearly broke a tooth when he bit on one to see how hard it was. He decided to tell no one at all, not even Katyusha, and before going home he dug a hole in a corner of the field. Into this hole he dropped the seeds one by one, counting them with loving care. Then he covered them up with the gray earth and put a white stone on top.

That night he was too excited to sleep, and he got up before dawn. He went straight to the field, removed the stone, took out one seed, and refilled the hole. Then he walked to Blansk with the precious nugget in his pocket. There he wasted no time in making his way to the shop of Boris Andreyevitch, the only tradesman and, guess what, the only rich man in the village. "Boris Andreyevitch," said Ivan Antonovitch, "how much is a nugget of gold worth?"

"That depends on the weight of it," was the reply. "But why should an old beggar like you wish to know such a thing?"

"That's why!" said the peasant, drawing the lump of gold from his pocket.

Boris Andreyevitch was astonished, but he took the seed and examined it carefully, rolling it back and forth between his fingers.

"Where did you get it?" he said at last.

"That's my business," said Ivan Antonovitch with a broad wink.

Boris Andreyevitch put the nugget on a small pair of scales, jotted down a lot of numbers on a scrap of paper, and finally said: "A hundred and twenty rubles."

"A hundred and fifty," snapped Ivan Antonovitch.

The old tradesman pretended to give this some thought, and said: "As you are a friend I will give you a hundred and thirty."

Ivan Antonovitch took the money and ran home.

"Katyusha," he cried in excitement. "Just look at this!" And he threw a handful of gold coins on the table.

"Holy Mother of God!" exclaimed his wife. "Where did you get all that money?"

"That's my business," said Ivan Antonovitch. But he could not keep his secret for long, and that same evening he took Katyusha to the sunflower field. There he lifted the stone and took out the seeds.

"Thirty-six!" he said. "And to think that yesterday there were thirty-seven!"

All night they discussed what was best to do, and decided that instead of selling any more of the seeds they would plant them. And so they did. They prepared a small field behind the house, and there they planted the seeds very carefully, in six rows of six.

The next day Ivan Antonovitch went to the village and asked

Boris Andreyevitch what was thirty-six times thirty-seven. "Next year we will have one thousand three hundred and thirty-two golden seeds," he told Katyusha when he got home. "You will have the richest husband in all Russia."

Then came the autumn rains and a long winter under the snow, and indeed it was the longest winter of their lives. When at last the spring came and the snow melted the two of them looked anxiously at the sodden black earth. And sure enough there were a few little shoots coming up.

In the months that followed the plants grew tall, and there was no doubt about it: they were real sunflowers. Every morning, as soon as they woke, Ivan and Katyusha would rush to the field. With bated breath they would gaze at the buds of the flowers, tight little fists holding all their immense riches. And one day, its face uplifted to the sky, the first sunflower opened its petals. It was a real sunflower . . . but alas, its seeds looked like ordinary sunflower seeds. Ivan Antonovitch pried one out with his fingernails. He crushed it between finger and thumb. Then all of a sudden he lost his head: one after another he tore open all the flowers, ripping off the petals. Inside them he found nothing but young, tender sunflower seeds. Then he grabbed his *amovar* and cut down the flowers in his rage. Katyusha looked on in horror, and when there was only one flower left she cried "Stop! Stop! Ivan Antonovitch, let us spare at least this one!" The man and his wife wept for a long time.

It was a sad, sad summer. And then the autumn came and the snow began to fall, but the sunflower they had spared did not lose its leaves, and its flower remained intact.

"Is this another miracle?" they wondered.

And one moonlit night they got their answer, for then they looked out and saw the flower gleaming like a golden crown. Running to the plant they found that the seeds which had been soft and gray in the daytime were now of shining gold. Ivan and Katyusha wept for joy and hugged each other and danced round and round the flower. "Better thirty-seven golden seeds than one thousand two . . . three . . ." In his excitement Ivan Antonovitch quite forgot the huge number the tradesman had taught him to say. ". . . Than a million golden weeds," put in Katyusha, who was always ready to coin a proverb.

So Ivan did not become the richest man in all Russia, but he was certainly the luckiest peasant in Blansk. He and Katyusha bought a

PL. XXVI Wo'swa, the bride of Pwa'ko

white horse and four big cows, and the next year they had a little daughter. They called her *Giraluna*.

The *Giraluna* appears as a mythological character in many legends of the North American Indians. The Xumi, who live in the Great Novaho Desert, are unfortunately dying out, but the few survivors of this magnificent tribe, which a couple of centuries ago still numbered two hundred thousand souls, keep the grand old poetic traditions alive. The episode we give below is part of the long epic poem "Ik'mia'ko," and among the protagonists is Wo'swa, the "moonflower." This particular legend is called "The Bride of Pwa'ko," and is one of the improvised dances for which the Xumi are famous but which now, deplorably enough, have sunk to the level of a tourist attraction. In the ballet Wo'swa wears a headdress PL. XXVI of white feathers decorated with seven shiny spheres, the "seeds" of her corona.

The Xumi language is one of the Kwo'na group, known for its phonetic richness and vast rhetorical potential, evident even to those who do not understand the meaning. In view of these qualities we give the episode both in the original (with a word-by-word translation underneath) and in a free rendering by Wallace M. Donovan of the Museum of Indian Traditions in Tucson, Arizona.

i'pwa'ko 'shi'opkuno
of Pwako   the bride

Pwa'ko'he si'ma'kwe te si'fushi i'yokwu imu'pnai'to mu' kwa'ma an'she
   Pwako   of Makwe and Fushi   hunt   Kwama

te'wu nau'sone a'nanamei i'ku mu'etto i'swe lu'too lu'konsiwa i'pwa'ko
   seven-color-tail   follow   Too buffalo   Pwako

mu'etto kwa i'kiusha na mu'kwa'ma 'na anwo chu'ettowe ti'la nap'kiu i'tu
   follow   night   Kwama   bat   distant

pwa'ko'ni ten'okwina tna te'hula ha twa mu'kwa'ma na che'ten'ta a'huyu
   Pwako   lose   help   moon

i'pwa'ko 'hle sha ma'kwe te kna si'fushi 'mi'tan'ta sha'mi'li i'si'wo'swa
   Pwako   of Makwe   and Fushi   moon   send   Woswa

kni wunau'kiashe tikiakia ekte'apkunan kia'wulwikia elth'ka kiu wo
   seven seeds   silver   skirt-serpents   roots

an'nochijan twa ne mu'etto ne'pwa'ko mu'etto'swi i'wo'swa tuwo lu'too kwe'i
   gazelle-leg   follow   Pwako   follow   Woswa   Too

a'wo'swa ne we'atina in'i'pwa'ko twesa ko'leho'li shi'wammina twe lu'konsi
   Woswa   black   Pwako   kill   most-big-buffalo

i'wo'swa tan'ta tse'ki a'shaw'li pwa'ko'na pish'le kwa u'liuna te tehula
   Woswa to moon return   Pwako   sleep   ask

ana tan'ta a'huyu i'pwan'chau te'opkuno i'pwa'ko na mu'etto kwa te'opkuno
   moon   from sky   young   Pwako   follow   young

i'pwa'ko 'ite'chu'nia ak'we pish'le anopwa'ko twa ko u'liuna te'opkuno
   Pwako   sleep   Pwako   ask   young

te'wo i'kwe'tenoka pa'yatamu tiu kwa tan'ta te'optekuna kwe i'pwa'ko
   Who are?   Payatamu   great feast   Pwako

a'pa'yatamu twe shi'optekunawi i'pwa'ko ushi'mu'etto twa iu'too
   Payatamu   bride   Pwako   lead me   Too

i'kiusha'ah twa pwa'ko ne pish'le twa i'kiusha zem'akwiwe a'pwa'ko
   darkness   Pwako   sleep   nights   twelve   Pwako

ha'ova i'pa'yatamu twa mu'etto a'pwa'ko ne u'she'liuna i'mu'tan'ta
   waken Payatamu   go   Pwako   call   moon

i'nu'pa'yatamu i'pwan'chau kwa te'wo'swa a'nota wo'swa shi'opkuno zwe
   Payatamu   from sky   Woswa   your bride

a'pwa'ko an'teyo kwa shi'wo'swa u'hluwa ne twa wunau'kiashe i'nu'chupachi
   Pwako   strike   Woswa   fall   seven seeds   birth

i'etto'chu lu'kon'siweko twa ko'leho'li twa a'nanamei te'winau'sone in'ta
   fox   rabbit   coyote   rattlesnake   wolf

intwaso ti'yi'ahalo tuwo konwasilu i'kwano'kio tzwe zem'akwiwe we'pachichuna
   gray turkey   eagle   cloud-rain   twelve   days.

The Bride of Pwa'ko (free rendering)

One day the young man Pwa'ko, son of Ma'kwe the tortoise and
Fushi the badger, went out hunting. Kwa'ma the bird with the seven-
colored tail approached him. Kwa'ma said to Pwa'ko: "If you follow
me I will lead you to the River Too where the buffalo go to drink."
Kwa'ma flew into the forest and Pwa'ko followed him. Night came.
In the darkness, the bright colors of Kwa'ma's tail could no longer
be seen, and he flew away like a bat. Pwa'ko had no guide now, so
he got lost. He asked the Moon to help him. "I am Pwa'ko," he said,

"son of Ma'kwe the tortoise and Fushi the badger." And the Moon sent him the great flower Wo'swa. Her seven seeds shone like silver and she wore a skirt of serpents and her roots were the legs of a gazelle. Wo'swa said to Pwa'ko, "Follow me. I will lead you to the River Too." Pwa'ko followed the great flower and at dawn they reached the valley where the river flowed. "Here is the River Too," said Wo'swa, who in the daylight had become quite black in color. Pwa'ko saw many buffalo. He took his bow and killed the biggest of them. "I must go back to the Moon," said the great flower, and vanished. Pwa'ko rested all that morning. Then he wished to return to his village, but when night fell he got lost again. Then he said to the Moon: "Send me Wo'swa, your flower." And lo, a young woman came down from the skies. "Follow me," she said. "I will lead you to your village." Pwa'ko followed the young woman and they walked for a long way. Then Pwa'ko said, "I am weary. I want to sleep." "Sleep then," said the woman. So Pwa'ko lay down and she stretched out beside him, and thus they slept for twelve days and nights. When Pwa'ko awoke the young woman was sleeping in his arms. He woke her, and asked, "Who are you?" And she answered, "I am Pa'yatamu, daughter of the Moon." And Pwa'ko said, "I want you as my bride." So they went to the village, and there was a great feast, and Pwa'ko married Pa'yatamu, daughter of the Moon.

One day Pwa'ko said to his bride, "I want to go back to the River Too, and you must lead me." So they went into the forest, and Pwa'ko followed his bride, who seemed to be made of silver from top to toe. Darkness came, and Pwa'ko said, "I am weary. Let us sleep." And they lay down and slept for twelve days and nights. But when Pwa'ko awoke his bride was no longer there. Then Pwa'ko called to the Moon: "O Moon, where is your daughter?" And the Moon said nothing, but down from the skies came the flower Wos'wa. "I want Pa'yatamu, my bride," said Pwa'ko. "I am your bride," said the flower. Then Pwa'ko was very angry, and he struck the Wo'swa. The seven silver seeds fell to the earth, and from them were born the fox, the rabbit, the coyote, the rattlesnake, the wolf, the gray turkey, and the eagle. A great cloud then covered the Moon, and it rained for twelve days and nights.

The seeds of the *Giraluna* also figure in the creation myths of certain African peoples. Typical of these is the Wombasa legend retold by Harold Wittens in his book *Under the Wombasa Sky*.[5]

The Sun and the Moon

Before they lived in the sky the Sun and the Moon dwelt on earth. They were peasants, and in their gardens they grew all the plants in the world. But one day the Sun made a flower that watched him wherever he went. When the Moon saw this flower she was envious, and she wanted it so much that during the night she stole it and planted it in her own garden. When day came the Sun could not see his favorite flower anywhere in his garden. He searched and searched; and during the night he found it in his neighbor's garden. There was a furious quarrel, and at last the Sun tore the plant from the earth so violently that its seeds flew right up into the sky. In this way were the stars created. The Moon climbed up on a cloud to get them back, but the cloud melted away and the Moon was left hanging in the sky. When day came the Sun planted another sun-flower in his garden, and the plant began to grow. It grew higher and higher, and at last grew so high that the Sun could not gather the seeds. So he climbed up the stalk. But when he reached the flower itself the stalk broke, and the Sun was also left hanging in the sky. Half the seeds fell into his garden and half into the Moon's garden, and so it came about that there are flowers of the day and flowers of the night.

Giraluna gigas

Giraluna gigas is so called on account of its unusual height, some specimens being among the tallest plants known to parallel botany. It normally measures over two meters in height, but can approach as much as four meters, as in the case of the biggest of the Lady Isobel Middleton group. This group is in fact perfectly representative of *G. gigas*. It consists of three plants which include all the morphological features of the variety, as well as a number of somewhat disconcerting anomalies that have not failed to produce some wild theories and bitter disputes in the worlds of botany and biology.

The sensational discovery of this huge group of flowers in the Tampala mountains provides a good example of the determination and personal sacrifice that often lie behind the terse, impassive language of scientific communication. Luckily for us this discovery was the subject of a monograph by Maessens, issued in 1972 by the London publisher George Allen Thomas, who was for many years the close friend and patron of the famous Belgian biologist whose recent death is a great loss to science.

It was not Maessens who first discovered the group. He was in fact told about it by his compatriot Paul van Berghen, who in turn had heard it mentioned during a visit he paid to the London head-quarters of the Royal Society for the Advancement of Parabotany. At a dinner given in his honor he met Sir Joseph Middleton, who had just returned from an expedition to the mountains of Tampala in northeastern India, an area particularly rich in parallel flora. The object of the expedition was to collect specimens of *Protorbis minor,* a kind of parallel mushroom which is quite commonly found under the huge *genensa* trees of the Landur forests, but which no one had ever managed to transport. Like many other parallel plants, *Protorbis minor* is no sooner touched than it dissolves into a pinch of grayish dust which certain tribes in the region mix with hallucinatory substances.

Sir Joseph told Van Berghen briefly about his expedition, and with tears in his eyes and a voice half strangled with emotion he recounted its tragic end. The details were passed on to Maessens a few months later in the course of a long recorded interview. He then compared this account with other testimony, and with several documents including the mysterious photographs taken by Marshall Norton, the news photographer and amateur parabotanist who accompanied the Middletons on their last adventure in India.

The expedition took place at the end of 1970, and consisted of Sir Joseph and Lady Isobel, Marshall Norton, and Patrick Hume, a chemical engineer who had invented a process whereby objects could be instantaneously enclosed in steophytic plastics. They halted for a while at the village of Banampur to organize the final phase of the expedition, and decided to spend the whole month of October in a strategically placed clearing in the Landur jungle. This clearing was duly marked on their map with the letter *M*.

On September 27, the small group left Banampur, along with a score of Amished porters laden with cases of camping equipment, scientific instruments, medicines, tinned food, and leathern bags full of drinking water.

The expedition reached point *M* right on time, on September 30. A few days were required to set up camp and organize the work program, but the prospects of success looked bright from the start.

Indeed as early as the second day Lady Isobel came across a group of *Protorbis minor* only a few yards from the camp, while at the third attempt Hume succeeded in enclosing one *in loco* within a block of steophytyrol fifteen by fifteen by fifteen centimeters. This is now the most perfect inclusion among the seven splendid specimens on view in the Birmingham Natural History Museum.

During the next three weeks the Middletons collected twenty-four *P. minor,* several woodland tweezers which Hume managed to implastify, and a small *Solea fortius* with an unusually fine series of protuberances. A discovery of quite exceptional interest was made in the neighborhood of some caves which showed signs of having been once inhabited. This was a unique group of stone figures of the Akda period, rather roughly carved, which depicted erotic scenes in which Yakanan and Drapanias copulated according to the Thamed rite, while holding in their hands small *Solea argentea* whose tips were in the form of a *lingam* and a *yoni*.[6]

It was Marshall Norton who by a curious concatenation of circumstances revealed the existence of the group of *Giraluna*. He personally recounted to Maessens all the events that led up to this spectacular discovery. For the day of October 21 the four of them had planned a reconnaissance on horseback toward the valley of the Andrapati, on the western side of the jungle, but at the last moment Lady Isobel said she was not feeling well and Sir Joseph decided to stay behind in camp to keep her company. Patrick Hume, who had had some technical troubles with his last inclusions, seized the chance offered by this sudden change of plans to look into the partial failure of his equipment. So it came about that Norton was left to do this particular trip alone. In his knapsack he packed a bottle of water, a box of biscuits, and a camera, and at about ten o'clock in the morning he rode off into the dark forest. Luckily the camera he took was a Japanese Dakon with a polyephymerol lens expressly designed and made for the chrommetric photography of parallel plants. Such a camera is capable of registering colored pigments that are invisible to the human eye.

Toward the end of the morning Norton arrived at the edge of the jungle, tied his horse to the branch of a centuries-old *genensa* tree, and sat down on a large boulder to admire the view of the wonderful valley where, according to the Pradahana, Shiva brought the sacred calf Nandi to graze in the spring.

A slight heat haze lay on the grass, and the surrounding moun-

Fig. 22 Marshall Norton's photograph of the Lady Isobel Middleton
group

tains were blue-violet. The River Andrapati (from Sanskrit *andrà*, lazy) lay like a heavy silver ribbon in the bottom of the broad valley. The immensity of the silence was accentuated by just a few sounds: the rustle of the *genensa* leaves as the horse tossed his head, the distant and sporadic cry of the hooting oopoopa, a native of the valley, which from time to time seemed to be answering the all-too-brief call of the cinnabar cricket.

Norton, enraptured by the beauty of the scene, was suddenly overcome by a great nostalgia. He had once described this feeling as "nostalgia for the present, which rouses the desire to fix forever the image of the moment that we are afraid of losing, and which in fact it seems we have already lost, while in reality it is still with us." Automatically he took the camera from his knapsack, raised it to set the focus . . . and realized that he had brought the wrong camera! But instinctively he went through the routine of taking a photograph, setting the focus, clicking the shutter, aware all the time that he was performing a series of useless gestures, a kind of empty ritual. But things did not turn out that way at all. The photo he took was in all probability the most important revelation of the whole expedition[7] (Fig. 22).

The next day he developed the roll of film in the portable but perfectly equipped darkroom that had been specially designed for the expedition by Seckton Brothers of London. He then found that, contrary to his expectations, the photograph was perfect, that the polyephymerol lens had acted as a polarizing filter and penetrated the haze, revealing details that had not been visible to the naked eye. By far the most obvious of these, right in the foreground, was the strange silhouette of a group of three enormous flowers. They could not have been more than ten yards from the spot where Norton had been sitting, and he was absolutely certain that he had not seen them.

Norton was so utterly astonished that in spite of the late hour he rushed to the Middletons' tent to wake them up and share the discovery with them. But he found the kerosene lamp still burning, while Sir Joseph and Hume, fully dressed, were sitting beside Lady Isobel's camp bed. She seemed to be deeply asleep.

Hume tiptoed out of the tent and told Norton that in the last few hours Lady Isobel's condition had suddenly worsened. She had a raging fever that neither antibiotics nor injections of pirianthro-pophyllin had had the least effect on.

After a sleepless night the party decided to return to Banampur, where there was a doctor who even spoke a few words of English. It was the end of the expedition. Lady Isobel lay for two days and nights in a coma before she died, her body covered with small purplish blotches, in the little white temple of Banampur dedicated to the paunchy elephant-god Ganesha and transformed for the occasion into an improvised hospital room. As if by some kind of presage the temple had been completely rewhitewashed, inside and out, just a few weeks earlier. Also whitewashed was the only piece of furniture in the place, an English Victorian chest of drawers which served as an altar. On this was the statue of the god himself, from whose trunk there now hung the hypodermoclysis which had been used in the last desperate attempt to save the poor woman's life.

As soon as they learned of the death of Lady Isobel, the men of Banampur built a pyre on the little beach formed by a loop in the Bahtra River, and at sunset six Amished youths bore her body down to the beach, wrapped in a white sheet and placed upon a board which they held high above their heads. When the body had been arranged upon the pyre the face was unveiled. Sir Joseph came forward with slow and heavy steps, and into his wife's crossed hands he pressed a little cube of plastic containing a *Protorbis minor.* By his own wish, he himself set light to the pyre.

A few days later the three men started on the return journey. With them, together with their priceless scientific finds, they carried a small canister containing the ashes of the great biologist. And one foggy January night Sir Joseph emptied the ashes into the Thames, in the knowledge that in some very distant place unknown to man the gray waters of the Thames would mingle with those of the sacred Ganges, and consecrate the memory of Lady Isobel.

Less moving, perhaps, but a better guarantee of immortality, was the ceremony which took place a few months later at the Society for the Advancement of Parabotany. On this occasion the group of *Giraluna* from the Andrapati Valley, the most important single discovery in parallel botany, was officially given the name of Lady Isobel Middleton group.

It was then that Maessens decided to study the three *Giraluna gigas* in person. On October 1, 1971, he started for the valley, stopping overnight in Banampur, where he visited the ill-fated temple. A piece of transparent plastic tubing was still hanging from the trunk of the elephant-god. He called on Dr. San to convey good wishes from Sir Joseph, and at dawn on October 2, accompanied by one of the young porters from the Middleton expedition and carefully following the itinerary mapped out for him by Norton, he set out on the long journey on horseback.

At sunset he reached the exact spot where the photographer had stopped, and sat on the very stone that had been described to him. He looked slowly around him, observing everything, and then he took out the now-famous photograph of the three *Giraluna,* which Norton had given him on the day of his departure from London. Everything fitted precisely. It was as if the landscape, creeping little by little into his consciousness, emerging like the image of a Polaroid photograph, came gradually to coincide with the picture he held in his hand. The vastness of the silence, also, suddenly deprived of the oopoopa's cry, was the very same that Norton had described with such loving exactitude. Maessens had planned to spend the night on that stone, beside the huge *genensa* tree to which he, like Norton before him, had tethered the horses. The young Amished squatted down at the foot of the tree and went to sleep at once, while the scientist sat and stared into the dark, waiting for the flowers to appear in the ambiguous light of the full moon.

He sat there for a long time. At last, at about two o'clock, and exactly where the photograph showed the silhouettes of the three plants, Maessens began to make out what at first was a barely discernible transparent shadow. Gradually, however, it took on more apparent substance, until the three flowers were perfectly visible.

In his little book *An Adventure in Parallel Botany,*[8] Maessens does not attempt to conceal the emotion he felt at that moment. "I stayed fixed to that stone as if I were myself a part of it, holding my breath for fear that the least noise or movement might break the charm and cancel the image that was in the process of forming." However, during the next two hours he was able to observe, measure and record at leisure everything he needed for a complete examination of the plants. It was when he finally decided to approach the group that their outlines seemed literally to melt away. They did not return, though Maessens remained on the spot until dawn, when he rode back exhausted to Banampur, with his notebook crammed with information. And here in brief are his observations.

The group consists of three flowers, two more or less equal in height (GA and GB) and one, GC, which is taller and somewhat different in form. GA and GB correspond in every detail to the description of *Giraluna gigas* previously given by Hydendorp. Their overall height is 2.85 meters, while their abundant *vetullae* are composed of about a hundred pendulants, of an average diameter of three centimeters. The lowest of these extend onto the ground and give a look of stability to the plants. The column, of which only the upper part was not hidden by the *avvulta,* has a diameter of about twenty centimeters and is flecked here and there with paramimetic bark. The coronas have no petals, but contain spherostills twenty-five millimeters in diameter which gleam with a yellowish metallic light vaguely reminiscent of brass. Flower GA has fifty-four spherostills, while to judge from the empty sockets in its corona flower GB is missing eighteen. The coronas are forty-eight centimeters in diameter and about eighteen centimeters thick. The angle between the corona and the column is about thirty-seven degrees.

Flower GC, both in the description given by Maessens and in Norton's original photograph, is in many respects different from the other two, and it is about this third flower that Hydendorp and other scientists have expressed doubts and reservations. According to Maessens it has an overall height of 3.60 meters, while the diameter of the column is less than that of the other two. The corona, which in proportion to its height ought to be well over fifty centimeters in diameter, is also smaller than those of the others, and is completely without "seeds." But the oddest feature of this flower is that it has no *avvulta.* There are a few very long pendulants which appear to have been crushed against the column and look like part of it. Although on a completely different scale, of course, they are a little like the dribbles of wax running down the sides of a candle. Maessens declares that beside its two sisters GC gives the impression of being dead. "But for the fact that we are concerned with a parallel plant," he writes, "I would have no hesitation in describing it as a dead plant. But there can be no death without life in the proper sense, and this cannot exist without the binomial time/substance (t/s). Now, as we know, parallel flora owes its entire nature to the absence of this binomial. It is therefore impossible that the tallest *Giraluna* of the Lady Isobel Middleton group could be a dead plant, or to put it another way, less living than the other two." For this author the plant only has individual features that are different, and one gets the impression that he is at pains to avoid the admission of an anomaly.

The publication of his study gave rise to a storm of conjectures and hypotheses. Some scholars, such as Giraudy, even hold the opinion that it is not a *Giraluna* at all, and perhaps not even a parallel plant. Deliberately ignoring the selenotropic existence of the plant, he maintains that chance alone has placed this flower beside two others of similar form. He thinks that it might even be a petrified sunflower of the post-Erocene era, and to support his view he points to the absence of spherostills and of any real pendulants to speak of. But he does not explain how a fossil plant, concrete and inert, can vanish at the approach of a man, as the testimony of Maessens expressly states. Hydendorp attacks Giraudy directly, and puts forward the simple hypothesis of an anomaly. He says that known examples of parallel plants are now sufficiently numerous to allow us to generalize and classify. But we have no guarantee that just beyond the limits of our knowledge there is not some completely baffling anomaly. "If we know of only two individuals of a species," says Hydendorp, "it is absurd to call one of them anomalous. If one of them were to be so called, which one would it be? But if we know of a thousand specimens an anomaly would become possible, and with ten thousand it would be probable."

But the basic question remains: Is the concept of anomaly possible in the case of parallel plants? In normal botany anomalies occur due to internal and external acts and events. Irregularities, fractures, or a wrong distribution of the anthrosomes in the Olsen chain can certainly cause changes of form (e.g. the four-leafed clover), of color (September rose), or of surface (scaly plane tree). It is the same for atmospheric agents, the actions of men and animals, and ecological changes of all kinds. But in the case of parallel plants, for the individual as well as for the species, formal characteristics are part of a special and constant mode of being themselves. If self-presentation is sufficient to justify the generalization that defines a species, then it will also justify those particular formal deviations that make each individual different from the others and recognizable as a separate entity.

In the specific case of flower GC of the Lady Isobel Middleton group we can only go on waiting for new discoveries and explanations. If Giraudy is wrong, and chance has not given us a petrified sunflower of the post-Erocene era beside two *Giraluna*, then the possibility remains that what we have is an anomalous specimen of a hitherto unknown plant similar to the *Giraluna*. Only the future can tell us for certain.

Giraluna minor

This is a miniature *Giraluna* that scarcely ever reaches, and never PL. XXVII exceeds, ten centimeters in height. Its habitat is the undergrowth, and particularly that of the creeping *Anaclea* that wreathes itself in a chaotic network around the knotty roots of the *genensa* and white *tarica* trees of tropical forests. Hidden in the darkness cast by this luxuriant vegetation, *G. minor* lives alone or in small groups, offering its tiny gleaming seeds to an invisible moon.

The flower has all the features of the common *Giraluna*. The differences lie entirely in its size and proportions. The column is little more than seven centimeters tall, and is enveloped right up to the corona in a small *avvulta* composed of hundreds of pendulants as fine as threads, which sink into the black moss at the base of the plant. The corona is proportionately quite large, its diameter often reaching half the total height of the plant, and it has a few rudimentary petals. The "seeds" are 2.5 millimeters, the same size as the ball bearings used for bicycles. They are shinier than the spherostills of the common *Giraluna*, resembling silver if they can be said to resemble anything. It is hard to understand how they can shine, buried as they are in the nearly total darkness of the undergrowth. They impress one as being full of an interior light, like a voice crying out in protest or desire.

Giraluna minor has sometimes been mistaken for *Protorbis minor*, which shares the same environmental conditions. The two plants resemble one another chiefly in size and in color, but *P. minor* does not have the spherostills which are such an outstanding feature of *G. minor*.

Hydendorp, to whom we owe the detailed description of these minuscule parallel flowers, has collected a good number of specimens, mostly set in plastic by the instantaneous *in loco* method developed by Hume for the preservation of *P. minor*. But his collection also includes a few rare specimens which have been transhabitated along with a sizable chunk of surrounding soil. Unfortunately the plants were taken to England a few months ago, and they are losing their substance so rapidly that it is feared they will soon cease to exist altogether. We may be left with the spherostills and the slice of original habitat, which will be useful to Hydendorp for his studies in parallel ecology, a branch of science of which he has only just sketched out the basic principles.[9]

G. *minor* has been found in nearly all the tropical rain forests of the world, from Brazil to New Anantolia. The Indians of the Rio Rojo attribute special powers to these minute nocturnal plants. They say that the young alligators that infest the river have an overwhelming craving for them, and as they wander further and further from the river in their frantic search, they eventually die of hunger and thirst. If it were not for this, according to the Indians, the river would be so overpopulated with alligators that it would become a solid mass of the creatures, a hard scaly road through the jungle.

PL. XXVII *Giraluna minor* in typical habitat

THE SOLEA

O f all the parallel plants of the Beta group, the *Solea* is the most complex in meanings and suggestions. Luckily for us it is also the one of which we have the largest number of detailed and convincing reconstructions, and by a process of repeated comparison we have gained a complete and exact knowledge of its morphology. The plant is extremely widespread in geographical distribution but is nevertheless fairly rare, and its origins are lost in the depths of time. Our knowledge therefore represents a considerable scientific feat.

Unlike other parallel plants, in which the relation between form and meaning is nearly always tenuous and often concealed, the *Solea* expresses its existential problems both directly and dramatically through its physical mode of being. Its nature and its morphology are inseparable, and they must be examined together in order to shed light, if possible, on their curious interdependence.

This is not easy, because if, on the one hand, the *Solea* expresses itself through its form, on the other, we are faced with the fact that what this form expresses is the complete inability of the plant to express itself. Paradoxically, its drive to self-assertion in the complex world of botany culminates in a language which, though rudimentary, in some way succeeds in embodying the frustrations arising from not being able to produce leaves, fruit, branches, flowers, and even a real set of roots. Spinder points out that this implies a form of "vegetal" imagination, obviously at a subconscious level, an awareness of its limitations, of what it lacks.

"But," queries Spinder, "what kind of self-image can a plant have?"[1] The gradual realization of a genetic program implies a

continual process of proposal and feedback, in constant reference to
a complete and final model. In the course of growth, any deviations
from this program have to be checked and corrected, while tempta-
tions have to be suppressed and anxieties soothed. And what is
more, we have to reckon with the environment, which imposes its
inexorable limitations from without. In normal botany, to bring
about the final "presence" of a plant, to perfect its individual mode
of being and its insertion into the general system of the biosphere,
this process is carried out constantly by all the organs, and by each
according to the function proper to it. In other words the genetic
program is the instrument by which the adult plant is formed, while
the model, like a kind of consciousness or self-image, is spread
throughout all the parts of the plant. If the *Solea* belonged to
normal botany, the cause of its pathetic arrested development might
well be found, like a phenomenon of degeneration, in hostile environmental conditions or in the defective execution of the genetic
program. But this would at the most justify an individual anomaly.
The play of natural forces, utterly devoted to survival, would not
permit the evolution of a variety that was permanently incomplete
or crippled. In all probability, mutations of a defensive type would
come into play to bestow on the plant a different normality and give
birth to a new species.

In parallel botany, any such theory, which implies an ontogeny
based on the processes of growth, must obviously be rejected from
the start. Its plants, motionless in time, are foreign to any type of
evolutionary change, including degeneration. They are as they are,
impervious to conditions alien to their own manner of being. What
seems to suggest an anomaly, such as the stunted excrescences of
the *Solea*, must in fact be an integral and permanent part of their
morphology.

All the same, Spinder thinks that even within the limits of a
closed and static system such as parallel botany, the forms of the
Solea might have some precise significance. He compares the plant
to a piece of sculpture. "Its meanings," he says, "do not reveal
themselves gradually, as with a normal plant in the course of
growth. They simply *are*, in the same way as the idea of a piece of
sculpture *is* in the mind of the artist who creates it. We can only
attempt to read them if we move, as it were, from the outside of
time to the inside, thus repeating—albeit in reverse—the imaginary
creative process."

For Spinder it would be absurd to attribute to the *Solea* impulses or frustrations that would imply an awareness of its own body. If by means of its rudimentary forms the body of the plant expresses its incompleteness, its inhibitions, refusals and failures, this does not for a moment imply that this can be attributed to interior exigencies or external conditioning. It is all part of a manner of being, not a manner of becoming. "It is," writes Spinder, "like a miracle never brought to a conclusion, suspended at the very climax of its perform-ance. The fascination and beauty of the melancholy *Solea* lie in the motionlessness of its becoming, frozen at the critical moment of evolution."

Jonathan Chase accepts Spinder's theory but refuses to be inhib-ited by it. "The theory put forward by our illustrious colleague," he writes in *The Journal of Parallel Botany*, "precludes any description that is not a cold mechanical inventory of the morphological prop-erties of the plant. Without saying so in so many words, as if he were afraid of the consequences, Spinder implies that the form of the *Solea* is purely symbolic, as its very presence is: 'Symbol of itself, like a piece of sculpture.' But in this case we must have the courage to give free rein to our imagination. If, as seems likely, we cannot attribute an interior life to the *Solea*, then we must invent one for it."

Chase, thus freed from what he calls "the absurdities of dumb man's speech," goes on to explain the meaning of the *Solea* in the following way:

"The *Solea* does not, like normal plants, possess an awareness of its own being and aspirations. It does not live, like its sisters in normal botany, constantly in reference to a genetically prescribed model. Rather, we think it is pervaded by a general sensuality, a state of expectation, forever aroused by the signals it receives but always disappointed by its own responses. The messages that arrive from the outside and touch the minuscule erogenous zones distrib-uted all over its body, these are the parallel equivalent of the normal genetic program. The messages propose leaves and stalks, buds and flowers, but the *Solea*, daughter of a different kingdom, is sadly forced to reject them all. Formless growths, bunches of swellings, branches that cling to its body like veins, and here and there a crippled wing of a leaf, the maimed prelude to a leaf-bearing branch —these are the melancholy proofs of its frustrated dream."

The *Solea* can now be recognized in many ancient legends of the most diverse ethnic groups, but in Western scientific literature direct references to it are few and of doubtful veracity. The first mention of "a plant which cannot ripen, standing upright in the bare soil of the [rubber] plantations" is found in Antonio Guerrero's classic volume.[2] In 1896 an explorer and raconteur from Dijon, a certain Jacques Pubiennes, wrote an article for the *Journal des Familles* in which he alluded to "a plant obscene on account of its phallic verticality," a phrase which cost him his job on the magazine and his secretaryship of the Société de la Découverte, which then had its headquarters in the capital city of snails and mustard.

In his book *Flora South of the Border* John Foreman, an American naturalist, tells of a journey through Patagonia in 1902.[3] In the course of it, he writes, "we saw strange plants in those woods: bare stems completely without leaves, and even without color, to which the Indians attribute supernatural powers. They say that these plants, which look like the knobbly walking sticks carried by old men, that in these parts are called *paarstoks*, have not been seen to grow or to die within living memory."

More recently the Dutch science magazine *Wetenschaap voor Iedereen* carried an article in which a certain Jan Van Handel claimed to have attempted to pick some plants, which from his description bear some resemblance to the *Solea*, in a rubber plantation in Sumatra. He writes that in spite of the frail appearance of the plants, and the fact that they had no roots, it proved impossible to tear them up. When the *dajaks* working in the plantation realized what he was doing they became extremely agitated, but when he tried to ask them about the plants they fell silent and refused to reply.

We are uncertain about the accuracy of these descriptions. The only *Solea* actually gathered, and the only one we have been able to subject to firsthand examination in spite of its state of semidisintegration, is the specimen of S. *fortius* brought back from Tampala by Sir Joseph Middleton. Many scientists, including Spinder, have expressed doubts as to its attribution, and these doubts are partly based on the fact that near the top of the plant there are two little branches, one of which bears what looks like a small trifid leaf.

The *Solea* is not a selenotropic plant like the *Giraluna* and, theoretically, ought to be perfectly visible. But, as in the case of all plants of the Beta group, our knowledge unfortunately cannot be based on direct observation. The descriptions we have quoted above

Fig. 23 Marcello Vanni, director of the Campora Laboratory

are probably ingenuous attempts to take myths and legends gathered in places where the *Solea* has existed and pass them off as firsthand information.

Luckily, however, and thanks to a grant from the Joachim Rosenbach Foundation of Milwaukee, the Laboratorio delle Campora has succeeded in putting together a dozen casts of reconstructions of *Solea* found in as many collections, museums, and labora- PL. XXVIII tories throughout the world. This small group of specimens has in the first place enabled students to establish analogies which in turn have formed the basis for a first rough morphology of the plant. By collaborating with other laboratories, Professor Vanni, the director of the Laboratorio delle Campora, has been able not only to enlarge his collection but also, thanks to the personal generosity of Mrs. Emily Rosenbach, to buy a few very rare and precious original reconstructions made *in loco*. The laboratory has prepared a general inventory, updated to December 1975, of all the *Solea* in museums, laboratories, and private collections. So far twenty-eight plants have been recorded, but it is to be hoped that after the publication of this little catalogue of Vanni's there will be news of many more plants, now hidden away in secret or inaccessible places.

The catalogue includes a drawing of a *Solea* based on the collation of all known specimens: a theoretical *Solea*, therefore, incorporating the average forms, proportions, and features of all the plants studied. Vanni points out that there was no particular difficulty involved in achieving this generalized model, given the marked similarity between one specimen and another. As far as size is concerned the *Solea* can be divided into two main groups, those about forty centimeters in height (*S. minor*) and those which measure about eighty centimeters (*S. major*). The two exceptions to this rule are the para-parasitic "tree-trunk" *Solea* in the Birmingham Museum and the great gilded *Solea* recorded by Woodby, both one hundred and fifty centimeters in height. The proportions remain more or less constant in all varieties, and they also have protuberances of similar shape. We have thus been able to gain a fair degree of knowledge of the most striking feature of the plant, which is the particular *rhythm* which seems to involve all its parts and which appears to be the same for all known specimens. For some time Spinder worked with Vanni at the Laboratorio delle Campora, and it is to him that we owe the interesting account of the calculations that led to the precise definition of this rhythm. These were based on the study of the "model" *Solea*, which is now known to science as *Solea Vannii*. They tend to confirm the insight of the great Malgueña, who when he first saw the drawing exclaimed: "Parece una falseta!" It was in fact the Spanish botanist who gave the plant its generic name, which he took from the flamenco *soleares*, which the Andalusian gypsies call *soleá*. As this type of song bears such a marked resemblance to the plant, a brief explanation of it might help us to understand the latter's basic structure.

"The *soleares*," writes the flamencologist Donado Malgueña, brother of the famous botanist, "is one of the matrices of gypsy music, a *cante jondo*. Its name is probably a gypsy translation of the Spanish word *soledades*, subsequently abbreviated to *soleá*. Like many other flamenco songs the *soleá* is slow and melancholy. The subject is nearly always desperation, the pangs of a love despised or betrayed or lost.

Es tu queré como er biento
y er mio como la piera
que no tiene mobimiento

Yo me voy a gorbé loco
porque una bina che tengo
la está bendemiando otro[4]*

---

* Your love is like the wind / and mine like stone / is without motion.
I feel I am going mad / because my vineyard / is harvested by another.

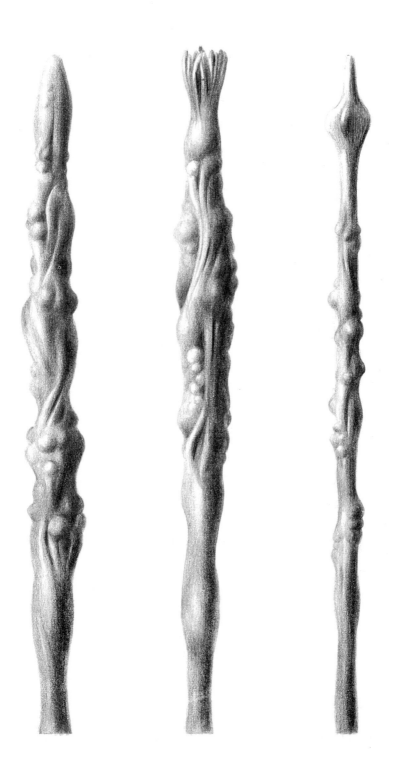

PL. XXVIII Casts of *Solea*

"The rhythm of the *soleá,* so hard to remember in spite of the heavy accentuation, is the very one that gives flamenco music that strange mixture of monotony and vitality which is so great a part of its charm. It consists of twelve beats with the accents on the third, sixth, eighth, tenth, and twelfth. In the course of the song and the dance these accents are often marked by clapping the hands (*palmas*) or tapping with the heels. The rhythm is called the *compás,* and it is performed on the guitar virtually as percussion. It is alternated with short musical phrases known as *falsetas,* which fill in the pauses between the verses of the song. The *compás* of the *falseta,* which is also strictly divided into twelve beats, is often disguised by frequent and prolonged *rubati.*"

The analogies between the topological rhythm of the protuberances on the *Solea,* which Spinder describes as "fructescences in the state of intention (*Urfruchten*)," and the time-rhythm of the flamenco *soleá* are so great that we are tempted to see a connection between the two. This is absurd, of course. Even so, the fact remains that S. *Vannii* can be read like a tablature. The distances between the twelve rudimentary "bunches" which are found in a spiral round the plant precisely reflect the sequence and proportions of the *compás:* 3-6-8-10-12. Within the bunches themselves the rhythm is much tighter, but it is the same. The single protuberances, which Jonathan Chase takes as representing a failed attempt at foliation, occur at intervals of six and twelve beats. All these excrescences are connected by long undulating filaments which alone or in groups envelop the plant and occasionally weave patterns. In the context of the rhythm of the *soleares* these would be *falsetas,* sometimes flowing and sad, at other times nervous and passionate.

In the summer of 1975 Vanni learned that the great guitarist Antonio de los Rietes was in nearby Siena for a series of lectures and concerts at the Accademia Chigiana, so he took this opportunity to invite him to the Laboratorio delle Campora. The *tocaor* from Jerez was immediately fascinated by the strange plants, and eagerly consented to take part in the experiments which Vanni proposed to him. These consisted in "translating" the spatial rhythms of the plants into musical terms. And so it came about that for some days de los Rietes was a guest at the laboratory, and there he "played" twelve *Solea* in the possession of the laboratory, as well as the theoretical *Vannii.* Recorded on tape, these *Solea* provide dramatic confirmation of the exactness of Spinder's calculations and the correctness of his interpretation. The guitarist himself was amazed at the result of the experiments, which constituted the first musical rendering of any plant, normal or parallel. The *Solea Vannii* now forms part of his concert repertoire of *cante jondo,* with the title "Camporanas." The music is characterized by long *ad lib* passages followed by brusque *a tempi,* which give it all the sensuality and melancholy so typical of the *soleares.* These *soleás* of de los Rietes do more than simply confirm the mathematical and topological insights of Spinder: they are proof of the analogies existing between the various specimens, the same analogies that made it possible to reconstruct *S. Vannii* in the first place.

There are only negligible differences between the rhythmic structures of the thirteen *soleares.* What gives each song its individual character is the tissue of the melodic lines, the length of the *falsetas,* the *rubati* derived from the shape of the veins, the protuberances, and the isolated amorphous growths. Nor should we forget the masterly execution of de los Rietes, that true *duende* of his that gives each piece an unmistakable stamp of its own.

A feature that might cause some surprise, on the other hand, is the traditional *rasgueado* which ends all these *soleares,* emphatic and violent but also in a sense stylized and banal, and not an PL. XXIX adequate expression of the tips of the plants which it is attempting to interpret. Evidently the Spanish guitarist did not succeed in reconciling the typically flamenco rhythmic and melodic coherence which all the plants share, with the spirit expressed by their tips, which vary from the festive baroque of No. 3 to the passionate simplicity of No. 19. His inability to find musical terms to generalize such marked differences spotlights a phenomenon which not even scientists have been able to understand. "How can it be," asks Spinder, "that a species showing such complex analogies between one specimen and another can permit each plant to express itself in a final statement of such utter individuality?" So far neither the Swiss biologist nor Professor Vanni has thought fit to advance any theories about this phenomenon, which is unique in parallel botany.

If we were to join up all the places on the globe where *Solea* have been seen and recorded, we would have a necklace as long as a meridian. Stories, myths, and legends referring directly or indirectly to the plant were once hard to find, but now they come to our attention quite frequently, and sometimes from the most unlikely places. This is largely due to the dedication and perseverance of

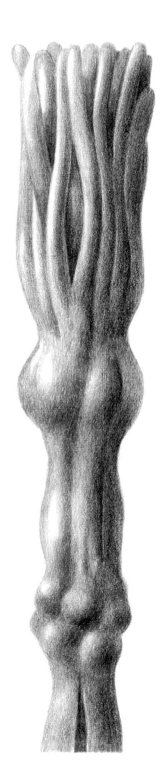

PL. XXIX Tips of *Solea*

Joseph Ascott, who has spent three years collecting and editing all available literature on the *Solea*. Ascott, who taught for many years at Columbia University, was in a position to use the vast network of informants and correspondents which previously enabled him to make New York the richest mine of anthropological documents in the world. Letters, telegrams, and even phone calls now quite regularly arrive at the Laboratorio delle Campora, from people who think they might possibly have some piece of information that in some way concerns the *Solea*. The bibliography of the plant is already fairly extensive, and from it we have chosen a number of legends that shed some light on the emotional impact the plant must have had on the vivid imaginations of peoples whose survival involved direct participation in the mysteries of nature.

The Silver-leafed Tchavo

In the village of Zibersk in Tarzistan they tell the story of the Silver-leafed Tchavo. Leo Lionni,[5] the famous author of books for children, heard it while traveling in Russia many years ago, and it was on this tale that he based one of his most celebrated fairy tales, "Tico and the Golden Wings." This American writer, who has been living for some years near Siena, Italy, is a frequent visitor at the Laboratorio delle Campora and a personal friend of the director, Professor Marcello Vanni. Thinking that he recognized the *Solea* in the generous shrub of the story, he communicated the fact to Joseph Ascott, who included the fable in his compendious bibliography of the plant.

Near Zibersk there are some grassy hills where the shepherds graze their sheep and goats in the summertime. On top of one of these hills there was a small thicket of Tchavo bushes, which grow to a height of about a meter and have straight smooth stems and very shiny leaves shaped rather like vine leaves.

In that thicket there was one Tchavo that never managed to grow like the others. It could only put out a few wretched buds that never burst into flower, but turned as hard as wood and seemed to die. There among all the flourishing leafy plants it looked as if it had been stripped bare by goats, although as everyone knows goats do not eat Tchavo leaves, which are as bitter as gall.

One day a poor shepherd, weary and sad, was sitting beside this miserable plant, this leafless warty stunted stem, and at a certain moment he burst into tears. "What am I to do?" he sobbed. "How am I to buy medicines for my little sick child?" Scarcely had he uttered the last word than one of the hard buds put forth a little twig bearing two silver leaves. The shepherd could barely believe his eyes. Very carefully he picked the two silver leaves, and to his astonishment he saw that their places were taken by two real leaves.

Some days later a young woman was sitting near the thicket, and the air was still and silent. Between one sigh and the next she said: "Ah, if only I had the money for a dowry I would get married!" Then she heard a sudden metallic rustle. A twig had sprouted from the stem, and it bore two silver leaves. Excitedly she plucked the leaves, and two real leaves grew in their stead. Then it was the turn of a poor peasant whose horse had just died: he too received two silver leaves. Then the same thing happened to a miller who had had all his flour eaten by rats. Each of them vowed to keep their good luck a secret, but even so the news of the Silver-leafed Tchavo did not take long to spread through the village. One day it reached the ears of Szabit, a rich moneylender. He hurried to the top of the hill as fast as his legs would carry him, and there he sat down beside the Tchavo and began to moan: "Alas, what am I to do, now that I am the poorest and most wretched man in all Russia?" Then he looked anxiously at the plant. Yes, a twig sprang out with two leaves, and then another, and another, until it was the tallest and finest plant in the whole thicket. But the leaves were not silver. They were real leaves, green and tender, that rustled in the wind.

Waa'ku-ni Creates Words

One of the most interesting legends of the Paraguayan Pa'nu'rà Indians, who live on the Rio Rojo, is the one which tells of the creation of words. It is a perfect example of the oral literature of South America, and was recorded by the French biologist Lamont-Paquit. It bears eloquent witness to the extraordinary gifts of wisdom, imagination and poetic instinct of this tribe, which ethnological textbooks seldom hesitate to describe as the most primitive in the entire American continent.

When Waa'ku-ni had created the earth he scratched ten furrows in it with his fingers and sowed the ten sounds of words. When spring came every furrow bloomed with red, blue and yellow flowers, and black flowers and white flowers; and Waa'ku-ni called them Ta-wa-tè. But between one furrow and the next Waa'ku-ni saw there were small bare stalks, without any flowers or leaves. He knew that he had not sown them, and he understood at once that they were the flowers of silence.

He sent his son Wo'ke down to the earth and told him: "Go and cultivate the Ta-wa-tè." So Wo'ke watered the flowers with the rain of his sweat, and the Ta-wa-tè grew tall and strong and put forth many buds. But the plants between one furrow and the next also grew, and one day, for fear that they should spread into the furrows, Wo-ke decided to pull them up. But every time he pulled up a flower of silence, one of the Ta-wa-tè lost its color. "What should I do?" he asked his father. "Leave them be," replied Waa'ku-ni, "for every flower of sound must dwell beside a flower of silence."

When the flowers were as tall as a man Waa'ku-ni said to Wo'ke, "Now make as many bundles of Ta-wa-tè as there are men and women on the earth, and give one bundle to each of them, and tell them to make words. Then make the same number of bundles of the flowers of silence, and give them to men and women, and tell them to make silence." And this Wo'ke did, and men and women were able to speak together, and to be silent and listen.

The Stake

The Patoná Indians who live on the south bank of the Rio de los Almas have a legend that was recorded by Randolph Reich and quoted in full in his *Botanical Psychogenesis*.[6]

On an island in the delta of the river there is a dense wood of larch trees. In this wood there lived a wicked white fox called Sipa. To persuade her not to eat their hens the Indians drove a stake in underneath the larches. Every evening they tied a toucan to this stake, and every night the fox came and carried off the toucan.

One morning the toucan was still there, and of the fox there was no sign. The Indians thought that the fox must surely be dead, so they waited for a day and then set the toucan free. But when they tried to remove the stake they found it was impossible. They asked the shaman of the village for advice, and he said: "Do not move the stake, for it contains the life of Sipa."

One night a hunter passing near the wood heard the sound of sobbing. He went nearer, and found it was the stake that was weeping. Then the shaman gathered together all the people of the village, and they sat round the stake and called upon the soul of the dead fox. Little by little the stake began to put forth buds and other small excrescences as hard as wood, and each time it put forth a bud it stopped weeping for a while. When it was all covered with buds and warts and excrescences the stake ceased weeping altogether.

Each year since then, the Indians have taken a dead toucan to the stake, and during the night the stake has devoured the toucan.

The Stone of Truth

The island of Taokeé is a great round flat totally bald pebble, one of the Baratonga Archipelago, which lies in the Pacific between latitudes 19° and 22° S and longitudes 161° and 165° W. In the few cracks to be found in its surface there is a very fine gray dust, so heavy that not even the March winds manage to shift it. When the sea is stormy, the spray mixes with this dust to form a kind of leaden plaster that little by little fills up the cracks. Within the next few decades the island will be completely without pores.

Long ago an Espak sparrow, probably lost from a sailing ship that was threading its way through the archipelago, flew over the island but did not alight there. However, it left on the island a few drops of its green excrement. At the end of the last century, Herman von Bockensteil, the only explorer who has dared to set foot on Taokeé, described these tiny spots as being a type of Klapaname lichen—"the only sign of life on the island."

For the Antonà, natives of the nearby islands, Taokeé is taboo. They do not come in close with their canoes even though they know that in the shadow of that great boulder there are millions of the little trementids that with their sweet flesh are a favorite prey of the fishermen of the islands.

The legend which explains this taboo was recorded by that un-

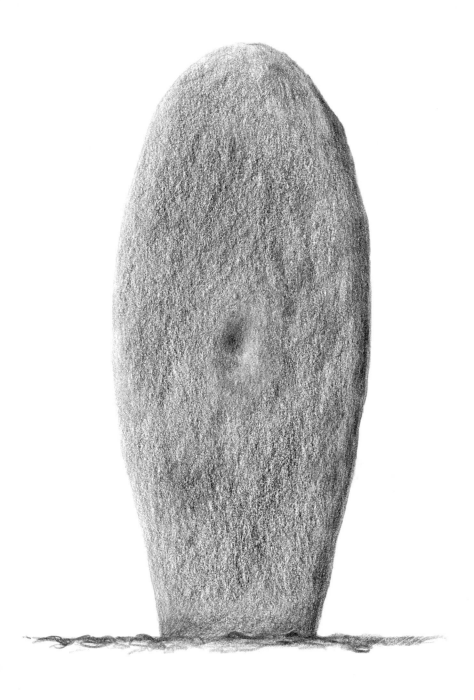

PL. XXX The great Stone of Tà

predictable but conscientious English reporter and helatologist Samuel Doncett, who paid a visit to the Baratonga group in the summer of 1907. He heard the story from a native of the nearby island of Tsa-wa, a certain Sep'a-nok. We have transcribed it from Doncett's original manuscript, now preserved in the library of the University of Hawaii.

Sep'a-nok put his lips to the great stone* in the midst of the PL. xxx village of Tsa-wa. The Antonà do this to show that what they are about to say is the absolute truth. This kind of kissing-oath is done on one's knees. In fact there is a smooth hollow in the stone about forty-five inches from the ground, a mute witness to the many thousands of tales which have been told by generations of natives.

Having taken his vow, Sep'a-nok told me the following story:

"The Great Disk O, father of all living creatures, cast eighteen handfuls of earth into the sea. Thus he made the eighteen Baratonga islands. With the clay that stuck to his fingers he made a flat round shape like a pat-là, and this he threw further away, and it became the island of Taokeé. But the soil of Taokeé was the dirt from the hand of the Great Disk, and for this reason it was more fertile than that of other islands in the group, so that many grasses and trees grew there.

"One day there was a great storm, and Taokeé turned upside down. The trees were left with their tops hanging down in the water, while the surface was covered with roots which soon died. The ants and the birds tried in vain to walk upside down with their legs in the air, but they were all drowned. When he saw all this, O said: "Ak se tikona," and the dead roots were turned to dust. Then he said: "A'se naré!" and the rains came down and the dust was turned to stone. And then he said: "Se-na nuaròa!" and a plant began to grow out of a hole that had been left in the very middle of the island. It had a mighty stem, but however hard it tried it could not put forth a single bud. Then it began to weep, and the Great Disk heard its weeping, and he said: "Sua nè poa!" and the plant went to sleep. In its sleep it dreamt a strange dream. It dreamt that it was surrounded by itself and was so numerous as to fill the whole island. When O saw the dream he said: "Anyone who wakes the

---

* This stone, which the Antonà call Tà, is now in the Anthropological Museum of Honolulu. [Author's note.]

flower from its dream will be eaten by the dragon He-Kà." Thus it happened that so as not to wake the plant the island became taboo, and no one has ever set foot there."

When he had finished I pointed to the distant island and asked Sep'a-nok, "But do you think that the plant is really there on the island?" The native looked at me in astonishment and said, "Can't you see that the island is covered with invisible plants?"

Even today the natives of Tsa-wa sometimes stop their Chevrolets near the spot where there once stood Tá, the stone of truth, and where there is now a traffic light. Kissing the iron pillar of the traffic light, they tell their children the legend of the *Solea* of Taokeé.

The Golden Spear of Tschwama

Bhinaswar is a city in the Indian State of Orissa, to the south of Calcutta. It is famous for its magnificent twelfth-century temples which rise like a thicket of red cacti prickly with statues and occupy almost the whole northeast corner of the city. Inside the temples there is a profusion of very fine bronze and stone statues of Ganesha, Vishnu, Ramesh, and all the other divinities of the promiscuous Olympus of the Hindus, while the great doors are embellished by *lingams* and *yonis* of all sizes and materials.

In front of one of the temples, the immensely sacred Tchhimbha, there is a stone platform bearing a very curious *lingam,* not smooth and stylized like the others but as thin as a stick and covered with warty knobs. Unlike the rounded hemisphere which tops the typical *lingam,* this comes to a rather misshapen point. If we were not aware of the phallic symbolism of anything that rises aggressively erect in the vicinity of a temple, we might be tempted to say that it was a petrified tree trunk, or at least a model of a tree that has lost all its branches. And we would not be far from the truth. This piece of sculpture has a strange history, which seems to originate in a legend in the Pradishana.

One day Dhana, the beautiful young wife of Prince Tschwama, went into the palace orchard to gather mangoes. The ripest fruit being on the highest branches, Dhana fetched her husband's golden spear to knock them to the ground. But she gave one blow that was so hard and awkward that the spear broke in two. Fearing her husband's anger, Dhana buried the pieces in the soft earth of the orchard.

When she returned to the palace, she found her husband there and he said: "Bring me my golden spear. A hundred warriors of Amhadur are besieging the city." Dhana could not confess to the truth, and she went into the orchard and wept bitterly. But where her tears fell, there the earth broke open and there grew a knobbly stem, as tall as a spear. Then Dhana heard a voice, and turning around in surprise she saw Lord Krishna in the branches of the great mango tree. "Take that stick," he told her. "It is the spear of Tschwama." Dhana took the stick from the earth and brought it to her husband. "Lord Krishna told me this is your spear," she said. Then Tschwama flew into a rage. "Do you take me for a fool?" he cried. And he seized the stick and began to beat his wife. But she felt nothing. Tschwama struck with all his force, but Dhana merely smiled and said: "I feel no pain." Then Tschwama was frightened, and dropped the stick on the ground.

Dhana picked up the stick and ran outside. On the green lawn before the palace Tschwama's horse was quietly grazing. Dhana leaped into the saddle and galloped off to the hill where the hundred warriors were waiting, brandishing the stick above her head. When the warriors saw a young woman coming against them with nothing but a knobbly stick they began to laugh, but Dhana attacked them one after the other, slaying twenty and putting the rest to flight. Then she returned to the orchard to put the stick back where she had found it.

But lo, in the earth where the stick had grown there now stood Tschwama's spear, shining and perfect. Dhana happily plucked it from the earth and put the stick in its place. Then she hurried to the palace. "Tschwama," she cried, "here is your spear!" But Tschwama had been informed of his wife's mighty feats, so he said: "Dhana, I do not want my spear. I want that stick."

Together they went into the orchard, but they found that the stick had grown a great web of branches and leaves, in the midst of which there hung two shining golden mangoes. There was a rustling in the leaves of the great mango tree, and looking up they saw Lord Krishna sitting among the branches. "Pick the mangoes of the Praham," he said. "But first, O Tschwama, you must break your spear, for you will need it no more."

Tschwama broke the spear and picked the fruit. Then he led

Dhana back into the palace and lay with her. Dhana bore a son and they called him Prahambhai. And every year on the twenty-seventh of July Tschwama and Dhana and Prahambhai would go into the palace orchard and pick the golden mangoes, while Lord Krishna played the flute among the branches.

Strangely enough we come across the dream of the *Solea* again, with a curious twist to it, in the memoirs of Bohm and Renner, who some years ago explored the Amazonian jungle not far from Manaus, and nearer still to the rubber plantations which at the end of the last century gave the city its moment of splendor and folly.

Surprised in the jungle by unusually torrential rains, the two Mormon geologists, who were prospecting for copper on behalf of Anaconda Copper Inc., took shelter in the communal hut of a village inhabited by Hanochucos Indians. There, in a pile of ritual objects, they noticed two baked clay tablets approximately one meter in length and covered with strange writing. They questioned the Indians about the meaning of this writing but met with a good deal of reticence. Finally, in return for a number of miraculous Polaroid photographs, they obtained an account of the two tablets.

Many generations ago, they said, a strange plant would appear from time to time among the rubber trees. This plant the Indians called the *oldikà,* and the strange thing was that it was able to dream. Being a plant, the *oldikà* could only dream about plants, but men being men they were able to dream the dream of the *oldikà.* One night an Indian who was working as watchman in a plantation went to sleep beside an *oldikà* and dreamt that he had picked the plant and rolled it along so as to leave an imprint in the light soil. When he awoke the plant was still beside him, perfectly intact, but the earth bore the imprint of the dream. Then the Indian was frightened, and went to the *omanashi* of the village and told him what had happened. The *omanashi* went with the Indian to the place, and there was the imprint of the dream, perfectly visible.

"You have dreamed the dream of the *oldikà,*" said the old man, "but it was not you who printed it on the earth."

"Who was it then?" asked the Indian.

The *omanashi* winked. "The *oldikà* is a sleepwalker."

Clearly the autonomous movement of the *Solea* can only be the fruit of folk imagination. Nevertheless, we are bound to admit that the clay tablets, now in the Natural History Museum in Salt Lake City, do quite definitely bear the imprint of a *Solea* and go some way to confirm the theory that at one time, before their parallelization, these plants grew in some abundance along the banks of the Amazon. There is no doubt whatever that someone did in fact roll a *Solea* along in the clay. What remains inexplicable is the fact that if the tablets are held upright the imprint appears convex or concave according to which direction the light is coming from. When lit from above the marks are protuberances in relief, like those of a real *Solea,* but when the light comes from below they look like ordinary imprints. This phenomenon, now known as oldicasis, has not been satisfactorily explained as yet.

The idea that a plant can dream, as happens in the legends of several ethnic groups in America, seems absolutely absurd in the light of our Western logic. But Solinez puts the following question: "If we admit that the pre-parallel *Solea* had some form of imagination (unimaginable for us but theoretically possible), even though at subcellular levels where the genetic memories lie in hiding, can we exclude the possibility of its having an active dream life?"[7]

THE SIGURYA

lthough between the earliest descriptions by Heraclitus (530 B.C.) and the recent reconstruction made by Maanen and Palladino there were only a few sporadic references to the *Sigurya*, we have today a fairly exhaustive knowledge of this plant, thanks chiefly to the patient research carried out by the Botanical Laboratory of Saragoza.

The name "Sigurya" was given to the plant by the celebrated flamencologist Donado Malgueña, brother of the botanist Juan Domingo Malgueña, who for many years was head of the Faculty of Biology at Saragoza University, and is still director of the botanical gardens of the city. It was in the Biblioteca Real, so rich in ancient scientific and pseudoscientific works, that Juan Malgueña first saw a picture of this plant, at that time completely unknown to him. He had come across a very rare volume, the *De Plantarum Mysteriis* of Paulus Aversus, a somewhat fantastical herbarium dating from the fifteenth century, and was looking slowly through the pages when one of the great woodcuts by Van Wittens caught his eye. It illustrated a number of monstrous hybrids, half animal and half plant, but also a few plants which appeared oddly plausible. He had no trouble in identifying *Laudanus umbrosus, Clariola foliata,* and *Opercus espinatus*, but however much he rummaged in his memory he found no trace of another plant, the one with the small aerial roots and the strange flower-fruit covered with warty knobs, which stood out so clearly against the background. This picture came back to mind the following morning, as he was walking down the great palm avenue toward the administration building of the Botanical Gardens.

It was then that Juan Malgueña determined to find out all he could about that disconcerting plant that looked so normal and yet clearly belonged to a different realm of botany. It is largely to the work of this elderly Spanish botanist that we owe our knowledge of the *Sigurya*.

Marcello Vanni, in an article published last year in *The Annals of Parallel Botany*, tells how he himself received confirmation of the ambiguous morphological character of the Sigurya, the thing that so powerfully struck the imagination of Malgueña.

"On the basis of indications provided by my assistant Paola Samonà, who is responsible for all the visual documentation in my laboratory," he writes, "I had done some rather detailed and particularly convincing pencil drawings of the *Sigurya*. My intention was to take them to Paris and present them to Juan Malgueña, who was to preside over the forthcoming Parallel Botany Conference at the Jardin des Plantes. As soon as I got to the French capital I phoned my old friend to make an appointment, and we decided to meet at the Café Flore in Saint-Germain-des-Prés, near where many of the delegates were lodged. When I got there, I found the old scientist with his wife and a number of mutual friends, including the well-known Italian photographer Ugo Mulas. In the course of the conversation it came out that it was Señora Malgueña's seventieth birthday. I took the opportunity to present to her the drawings I had brought for her husband.

"Everyone present studied the drawings with care, and during the animated conversation that inevitably followed, Ugo Mulas left the table unnoticed. We only realized he had been absent about ten minutes later when he reappeared with a huge bunch of flowers which he presented to the old lady. There was clapping and cries of "Buen compleaño," which a group of young American tourists joined in, singing "Happy Birthday, dear Señora." In the general uproar, Mulas handed me a small package wrapped in tissue paper, looking at me meanwhile with a mysterious air. I opened the package eagerly and could scarcely believe my eyes: it was a dried flower that lay there, or perhaps a fruit, practically identical to that of the *Sigurya* in my drawings. The photographer explained that when he had seen it at the florist's where he had bought the flowers for Señora Malgueña he had been as thunderstruck as I was.

"After the little impromptu party we all went off to the florist's in the hope of finding other specimens of this strange flower-fruit. But the shopkeeper said he had no more of them. When asked the name of the plant he searched for a long time through old brochures and catalogues, but without success.

"Juan Malgueña took on the job of finding out what he could at the small library attached to the Jardin des Plantes, and when the conference opened the following day, he handed me a sheet of paper covered with notes. The specimen, the notes said, seemed to be the fruit of *Santilana panamensis,* one of the family of Felinotenis, a native of Panama that now only survives in a few of the Caribbean islands. The islanders apparently dry the fruits and sell them to dealers as decorative plants. It seemed to be by no means a rare plant, but strangely enough, it was unknown to any of us, botanists and biologists alike. While the flower-*cum*-fruit was very like that of the *Sigurya,* the rest of the plant had really quite different features. The *Santilana* is a plant with an endinodal stem of medium height, bearing large coriolated leaves. The root is a rhizome that spreads under the ground and produces an average of ten individual plants."

The Greek philosopher Heraclitus, founder of the school of Gynos which flourished in the sixth century B.C. and inventor of the strobological theory of language, was the first to give a reasonably detailed description of the *Sigurya,* which he called *Gynopsa.* His observations are available to us only in a rather debased Latin translation, completely lacking in his own incisive style.* The translation describes the *Sigurya* as "a plant with a flower that looks like a head with nothing but noses wearing a skirt like the tasseled ones worn by the vestals of the oracle of Markos." Of the size of the plant the translation says, "It is as tall as my ten-year-old son Demoklitus; a goat would have to stretch its neck to eat the fruit of it."

Heraclitus confuses the fruit with the flower, but gives a surprisingly accurate description of an encounter with the plant. More banal, but certainly clearer, is the description by Maanen and Palladino which accompanies the reconstruction in wood made by the two sculptor-scientists. Maanen and Palladino were already cele-

---

* James Fadden has made an attempt at reconstructing the bizarre and almost incomprehensible manner of the lost original text, which is typical of the strobological school and has suggested to Burroughs and other modern writers the cross-cutting techniques popular in contemporary poetry and fiction.

Here is the breathtaking opening:

"In Tha (beyond) mos (Es-tor) Demoklitus (Theo-the) and I gath-olive-ered Demo (Theo) klitus (the-be) called. Plant (ant-plan)t skirt Mar vest (kos) tal he- (noses noses no) (ses) -ad."

brated for the wooden models in the Oosterman Museum in Nyme-
gen. These are vastly magnified models of the most minute details
of normal plants, such as the pistil and the chromostene of *Colido-
tima*. Particularly well known is their model of a colony of cells of
Folia antrax enlarged 1500 times and clearly demonstrating the
tendencies of perimetric cells in the process of directional and selective growth. It was Juan Malgueña who aroused their interest in
parallel plants, and particularly in the *Sigurya* to which he himself
had devoted so much intense effort. The truly wonderful model of
Sigurya barbulata which stands in a glass case in the center of the
huge entrance hall of the Botany Laboratory at Saragoza is the work
of these two. It has become a kind of touchstone for all descriptions
of the plant, even though it has a few features outside the usual run
of things.

The model is life-size and reproduces a *Sigurya barbulata* that
had actually been studied and reconstructed by Malgueña, who
worked in close collaboration with Maanen and Palladino. Visitors
to the laboratory are presented with a small brochure containing the
following information:

> At present we know of six kinds of *Sigurya*, but over the next
> few years this number is expected to grow considerably. In
> order of size these varieties are: *S. gigas grandiceps* (Big-
> headed *Sigurya*); *S. montalbana; S. barbulata; S. afrocarpus*
> (Dark-fruited *Sigurya*); *S. microthele* (Small-nippled *Sigurya*);
> *S. minima*. Peter Foreman has furnished evidence of an aquatic
> variety called *S. natans*, found in Ottogonia.

The Saragoza *S. barbulata* was the first to be found in the West. PL. XXXI
As it stands, the plant is sixty-two centimeters tall. Malgueña calculated that if one straightened out the curve in the stem which bears
the cephalocarpus, the height would increase by seventeen centimeters. The stem is known as the "corpus" and bears some resemblance to the column of the *Giraluna*, though it is far more slender.
It has three rings of aerial roots which Malgueña calls *barbules* and
which are superimposed upon each other without any apparent
order. They are at most four centimeters in diameter, and are the
equivalent of the pendulants of the *Giraluna*. The corpus is twenty-
two centimeters in diameter at the base and three centimeters at the
highest point, where it curves downward and holds the cephalocarpus.

This cephalocarpus is the most characteristic feature of the *Sigurya*. It is a kind of fruit, totally irregular in shape, with protuberances of various lengths sprouting from it all around. There are conflicting theories concerning the nature of the cephalocarpus, which so strongly resembles the fruit of the *Santilana panamensis*. Malgueña refuses to consider it the parallel equivalent of a real fruit, although it has all the appearance of a fruit. While they were working on the reconstruction, his colleagues referred to it as the "head," and this was why Malgueña coined the term *cephalocarpus*. For him, however, the cephalocarpus is the plant itself, the remainder (the corpus) being no more than a much-needed support, like the base of a lamp. His theory receives some measure of confirmation from the existence of two specimens of *Sigurya afrocarpus*, where the corpus is altogether lacking, and of a S. *minima* discovered by John Harpers near Opanò on Venderas Island. In this latter case there is practically no stem.

Olaf Rasmussen, director of the Parabotanical Research Center at Omloë, disagreed with Malgueña's theory in the paper he read to the Copenhagen Conference. According to Rasmussen there is in the whole of botany no fruit or parafruit that is not in some way supported. Even *Protorbis minor,* he observes, has a base which serves to keep the most expressive part of the plant off the ground. For Rasmussen the specimens mentioned by Malgueña are only fragments of broken plants, although the breakage may have occurred either before or after parallelization. In an open letter published in the Bulletin of the Omloë Center, he urges his colleagues in Ghana to go back to the Tarno River where the specimens of *Sigurya afrocarpus* were found and to search there for the missing stems.

The unfortunate fact is that information from distant countries, especially that regarding plants of the Beta group such as the *Sigurya*, is nearly always fragmentary and inexact, as well as being at least secondhand. But apart from the meticulously faithful reconstruction made by Malgueña and his colleagues, we also have information which has reached us due to the generosity of Ricardo Martínez, one of the archeologists responsible for the recent excavations near Oaxaca, Mexico, not far from the site of the famous Tomb No. 7 at Monte Alban. In a little book called *Homage to Gutiérrez,* Martínez tells how he discovered a large black clay vessel decorated with Mixtec graffiti and containing old weapons. It stood

PL. XXXI *Sigurya barbulata*

in the narrow entrance to the underground corridor which he supposed would lead him to the central chamber of Pyramid No. 3 ("la Desnuda"). It was thought at first that the weapons were Mixtec arms of an unknown type, so the discovery created quite a sensation. But closer study soon put things in their proper perspective. Carbon tests showed Martínez that while the vessel was certainly Mixtec the weapons had been inserted later, probably to hide them. But what was a disappointment to the archeologist was a triumph for Professor Pedro Gutiérrez, the aged honorary director of the School of Botany at Vera Cruz. Suffering from a serious lung condition, he happened to be on holiday at the same hotel, the Marques del Valle, where Martínez was staying. The two scientists had known each other for some years, and in the evening they would meet on the terrace of the hotel, overlooking the tree-lined square. From the *art nouveau* bandstand in the square a group of *mariachos* would scare clouds of birds from the trees with each blast of the trumpet, while the Olted boys would play hide-and-seek around the trunks of the huge flowering jacarandas. It was on one of those magical evenings, when the sunset seemed to have already lost the first of its gold-vermilion to the oncoming darkness, that Martínez told Gutiérrez of his perplexity about the weapons he had found in "la Desnuda." He invited his friend to visit the site.

The following day Gutiérrez went out to Monte Alban, where the archeologist lifted the lid of a huge vessel to show him its disappointing contents. There were fighting machetes, spear blades engraved with stylized symbols, and other metal objects. And on top of the pile, attached to a narrow, worn leather belt, was a bizarre shape vaguely resembling a medieval mace. It was indeed an unusual thing, apparently slightly rusty, and Gutiérrez picked it up with the greatest care in spite of his evident excitement. The shape was organic, a little larger than a clenched fist, and completely covered with large knobs and tentacle-like protuberances of various shapes and sizes. These looked like fingers, or small pendulants, and there were about thirty of them. For Gutiérrez there was no doubt about it: the shape was that of the cephalocarpus of the *Sigurya*.

The old botanist's mind was besieged by a swarm of questions. Was it a coincidence? Or a metallized plant? A copy? A fossil? He had no difficulty in persuading his friend to give him the loan of the object, which he took back with him to his room at the Marques del

Valle. There, writes Martínez, "Gutiérrez remained incommunicado
for three days and nights. In the evening of the fourth day he
appeared at the bar in time for an aperitif before dinner, sprightly
and elegant in his white linen suit. Accompanied by the languorous
crooning of a duo engaged for the week by the tourist board of
Oaxaca, he there and then confirmed to me his first intuition about
the weird thing among the weapons." He gave Martínez a few pages
covered with notes and diagrams, all of which the archeologist
reproduced in facsimile to illustrate his booklet. Gutiérrez died in
Vera Cruz only a few weeks after his return from Oaxaca.

With the Saragoza reconstruction, the account given by Martínez
and the notes and diagrams provided by Gutiérrez constitute the
most complete and scholarly documentation of the *Sigurya*. There is
no longer any doubt that the cephalocarpus at Monte Alban is a
complete plant, and is therefore evidence in support of the theory
advanced by Malgueña. With the plant on the ground, as it is shown
in a photo taken by Martínez, the protuberances look rather like
stunted pendulants, some of which act like normal roots and attach
the plant to the ground, while others look like the arms of a blind
man groping desperately in the air for some nonexistent handhold.
What neither Gutiérrez nor Malgueña were able to explain satisfactorily is the metallic nature of the plant. In his notes, Gutiérrez
speaks of a process similar to that which petrified the trees in the
Yosemite National Park in California. This point of view would
imply a process aimed at rendering permanent a plant which, being
parallel, was permanent in the first place. It would also imply the
transformation, by an infinitely slow process of mineralization, of a
nonmaterial object existing outside of time. None of this is compat-
ible with the theoretical premises of parallel botany. More plausible
is the theory put forward by Van der Haan, according to which the
Sigurya of Monte Alban is a concretion formed from the imprint of
a real plant. This concretion occurred, according to the theory, in
the course of an extremely violent earthquake that caused the
greater part of the ancient acropolis to collapse. This earthquake
also cracked open the surface of the earth over large areas, allowing
the escape of gases and liquids so hot that they were capable of
melting the ferrous minerals in which the zone abounds.

The *Sigurya* of Monte Alban is now on display in the little
museum of Oaxaca, together with the precious finds from Tomb No.
7. An enterprising jeweler in Taxco has made a tiny copy of it, and

this can be bought as a souvenir in the shops beneath the arcade in the picturesque little square.

Malgueña, seizing on the evidence that the *Sigurya* of Monte Alban was a complete plant, and therefore neither fruit nor flower, carries his theory to its logical conclusions. He declares that the corpus is a deceit, an extreme but authentic example of paramimesis. The furious debates on the nature of that ambiguous organ, still raging in the botanical world, are perhaps the best possible support for his point of view. After all, paramimesis has no other object than to create doubt and confusion, and to protect a plant that even an excess of zeal or love of science might easily destroy.

Sigurya natans, simply because it is an aquatic plant, does not fit <small>PL. xxxii</small> into the normal categories of parallel botany. Only two specimens are known. The more famous of these is the one described by a certain Jacopo della Barcaccia, who was with Magellan on his last great voyage, in a letter to his wife Dorotea. This priceless document was discovered in Padua by the historian Tschbersky. The other known specimen is the copy in wood made, like the *S. barbulata* of Saragoza, by the sculptor-scientists Maanen and Palladino.

As regards the *S. natans* described by Barcaccia, we need only reprint some part of the long letter mentioned above:

> In those crystal-clear waters [of a lake on one of the Termadores Islands] there are crabs with antlers like stags and many fishes covered with feathers like birds, and also eels as long as boats whose scales were like golden ducats; and men say they are mortally dangerous. There are also enormous turtles which the men who live on the water mount like horses, and so ride from one island to another. And in the trees there are fisher-birds with beaks as long as swords, that are a wonder to set eyes upon, and others that sing so sweetly that Fra Simone himself would envy them [Fra Simone was a composer and organist at the church of Santa Teresa in Padua].
>
> Also in those waters are strange fruits with fingers, and no man can touch them, for at a touch they dissolve and melt away, and the young men are forbidden to look at them, for they fade from the sight and make themselves invisible. This fruit they call *panalà,* and it filled me with wonder, for it is

neither tree nor shrub nor yet a flower, but it floats on the water and the roots hang down from it. It is as black as the ink of a squid, and when darkness creeps over the water all the men and women of the island pray to this fruit as if it were the *os sacrum* of St. Barnaby himself.

Tschbersky thinks that Barcaccia's descriptions have a lot in common with the more succinct annotations of Pigafetta, Magellan's official historian. He writes: "There is no doubt in my mind that the magic fruit floating in the lake on that island was indeed a *Sigurya natans*. Many other travelers who have followed Magellan's route to the Termadores Islands have confirmed Barcaccia's account. The *panalà* was indubitably the ultimate form of a group of plants which one after the other have vanished at the fatal touch of man."

The model carved by Maanen and Palladino is only partly based on the descriptions left us by the travelers of the sixteenth to eighteenth centuries. The information gathered by the missionary Father Beaulant, a keen student of flora who lived for many years among the natives of the Termadores, provided the sculptors with sufficient data for them to complete the work of reconstruction in the greatest detail. This model is also on show at the Botanical Laboratory of Saragoza, and thanks to the inclusion of the aquatic pendulants in a large block of Plexiglas it gives a really convincing feeling of reality. The pendulants are long and thin compared with those of the *Sigurya erecta* (this is Malgueña's name for the plant with the stem), and in river water they must undulate like the native algae of the stream. The theory that the *Sigurya* originated in the many little streams that come down from the mountains and feed the inland lakes of the islands is supported by some of the prayers in which the natives invoke the spirit of the *panalà* to stem the fury of the torrents in spring. Maanen, perhaps following some unofficial statement of Malgueña's, suggests that the *panalà* of the Termadores might be the mother of all *Siguryas*. In common with all the organisms of our planet, the *Sigurya* would therefore have distant aquatic ancestors, and the *S. erecta* variety would represent a more recent development, a second phase of parallelization.

PL. XXXII *Sigurya natans*

PART THREE

EPILOGUE

THE GIFT OF THAUMAS

For some years past the Swiss biologist Max Spinder has been going for his summer holidays to Emplos, a cluster of little white houses on the edge of the grounds of the Hotel Peleponnesus, high on the cliffs of Cape Antonosìas. Last summer, while walking under the centuries-old pines, he met the American archaeologist John Harris Altenhower, who was at Emplos to study the nearby ruins of the Temple of Kanos with a view to deciding whether or not to involve the University of Cranstone in an intensive excavation program. The two men, who share a love of Greece and a passion for exploring unknown worlds, soon became fast friends.

One morning they decided to go for a walk along the cliffs, following a path that wound between pines and myrtles until some three kilometers from Emplos it reached the snow-white fragments of the Temple of Kanos, scattered in the underbrush beneath the purifying sun. They talked of their work, and Spinder was deploring the incredulity with which the scientific establishment, and even some of his own colleagues, had greeted the news of facts that were as yet inexplicable but which he had proved experimentally. When they were a stone's throw from the ruins, Altenhower broke in on him to observe that more than two thousand years before, in that very place where they were walking, Heraclitus and Theaetetus had carried on the famous dialogue immortalized by Plato.

Taking his friend by the arm, as if he wished to re-create the scene, with an ironically theatrical gesture he declaimed the key sentence of the dialogue: "But if you, O Theaetetus, were to see among the myrtles a berry as white as a pearl and as cubic as a die, would you reject it with contempt and disgust as a horrible whim of nature, or would you pick it with joy and gratitude, as a divine gift

from Thaumas?"[1] It was then that Spinder brusquely shook off
Altenhower's hand from his arm, left the path and worked his way
laboriously through the dense undergrowth until he reached a large
piece of white marble, maybe a section of column, that lay almost
entirely hidden from view about ten meters from the path. There he
stopped, bent down and, almost overcome with emotion, called out
to his friend. When Altenhower joined him, anxious to know what
on earth was happening, Spinder pointed to two strange black
plants no more than twenty centimeters tall that stood upright like
little bronze statues in the midst of a minute clearing, a small circle
of bare earth amongst the prickly scrub.

There was a slight sea-breeze laden with the scents of seaweed
and thyme, so that the longest branches swayed on the pine trees
and the leaves fluttered on the bushes; but the two little plants
remained perfectly motionless, throwing a brightly colored and ex-
traordinarily luminous shadow on the burnt clay soil. It was as
though the sun's rays had passed miraculously through them as
through a prism, casting on the ground not a shadow but a rainbow.

The two men were overcome with astonishment, and for a while
they stood there staring at the sight in helpless speechlessness.
Spinder knew from experience that the plants would dissolve into
dust at the first touch, so he decided to return to Emplos and fetch
photographic equipment.

Unable to tear their thoughts away from the amazing vision they
had just witnessed, they both walked in silence. Suddenly Altenhower stopped dead. What extraordinary intuition, he asked, had
led Spinder to those plants, which from the path had been completely hidden. The biologist smiled and said: "I'm flattered by your
high opinion of my powers, but at the same time rather surprised by
your ingenuousness. You must surely know that there was nothing
miraculous about it. The water-diviner believes in the movements of
the rod, but the truth is that without knowing it he has an exceptionally sensitive reaction to certain natural things: colors, smells,
kinds of earth, the shapes of plants, all things that derive the
ultimate subtleties of their nature from the presence of water under
the soil. Like a frog with an instinctive perception of a pond some
miles away, he unconsciously distinguishes differences of shade and
size which would not be perceptible to us. And the same holds true
for the archeologist who 'has an intuition' of a buried temple under
a perfectly ordinary ploughed field, and the botanist who 'has an
intuition' of the presence of a parallel flower in the midst of a

thousand normal plants. They both read signs which little by little, through the continual habit of specialized observation, build up in the deepest levels of the memory. There they lie in readiness for the time when a particular combination of automatic analogies will call up images long forgotten and now remembered with instant clarity."

The discovery on the cliffs of Cape Antonosìas of the two *Parensae parumbrosae*, which Spinder, thanks to a brand-new process, was able to transhabitate with complete success to his laboratory at Hemmungen, was announced to the public in the latest issue of *The American Botanist*. It was the first time that the authoritative organ of the American Botany Association, which traditionally interests itself only in normal botany, had really opened its columns to a phenomenon of parallel botany. The evidence of Altenhower on the circumstances of the find, the description of the plants themselves and, above all, the phenomenon of the colored pseudoshadow which was perfectly visible in Spinder's photographs—all these aroused a good deal of sensation in scientific circles. Even today, some months after the news broke, the media are still devoting a lot of time and space to the event.

One of the first newspapers to take up the story was the Greek daily *Omonia*, which interviewed Professor Spyros Rodokanakis, Professor of Botany at Athens University. This veteran botanist is well known to the Athenians for his provocative attacks on what he calls "the invasion of reason." A few years ago his vitriolic sarcasm did not even spare the colonels, who for some reason best known to themselves chose to turn a blind eye to the violent attack on their regime which the professor launched from the pages of *Botanika*.

But the furious arguments and controversies carried on by Rodokanakis often close more doors than they open. He often unwittingly becomes the mouthpiece of those who, in the name of tradition, wisdom, and a kind of freedom that is never very well defined, obstinately refuse to leave the murky vapors of their own mental status quo. And so it was on the occasion of the short interview which he gave to the Athenian daily.

"It is fashionable," he said, "to stigmatize the mass media for the devilish way in which they create false needs and consequently contribute to the spread of manic consumption. But if labor-saving electric appliances and the small family car can atrophy our muscles, there are in my opinion far graver, more real and more imminent dangers threatening the survival of man. The so-called hidden persuaders are merely witless shopkeepers compared with those who in the name of cultural and scientific information pollute our minds and intelligences with ideas that could have no other purpose than to put an end to our already frail ability to tell perception from fantasy, reality from fiction, and truth from falsehood. These gentlemen have cynically sold us telepathy, alpha-rays, flying saucers, mental deconcentration, acupuncture, the Loch Ness monster, forks bent by willpower and the Black Box. These ghost hunters in nonexistent laboratories have now, it seems, discovered in the vegetable kingdom those anthropomorphic qualities which man himself is rapidly losing: the ability to feel joy and sorrow, a real love of the arts, a hatred of tyranny and even the use of a comprehensible language. We are told that we may safely and confidently engage a saxifrage to spy on our unfaithful spouse. They encourage us to play the tango and the kalamatianó to make roses grow more voluptuous and perfumed. They suggest we should recite the poems of Verlaine to straighten a wilting aspidistra in the waiting room of a Parisian dentist. And they assure us that while the voice of Gigliola Cinquetti weakens geraniums, that of Renata Tebaldi stiffens their stems.

"And now this glorious literature of fiction and fantasy has been enriched by a new masterpiece: among the sacred ruins of Kanos, where Heraclitus himself meditated, they have discovered a "parallel" plant as black as ink, that throws a shadow as bright and many-colored as the windows of Notre Dame. It will not be long before we hear that a cyclamen has been proclaimed Rector of the University of Athens."

But in the fury of his rancor the veteran botanist lumps together the absurd with the possible, madness with reason, good with ill. His mental inertia leads him to express a mere hotchpotch of refusals and denials, when a more open attitude, a calmer optimism, a more generous confidence in others, would certainly have rewarded him with unsuspected creative happiness. Though perhaps we can scarcely be surprised if the revelation of a parallel flora, splendidly enigmatic in character, has given rise to incredulity, skepticism and, on occasion, open hostility on the part of those who with blind bureaucratic resignation go on cultivating the old common-or-garden plants in their common gardens. We have to admit that in the wake of a perfectly understandable alarm, the fascination of mysterious and ambiguous organisms suddenly wrenched from the deep shadows of the jungle and from the mists of legendary valleys has led at times to the hasty formulation of exotic theories and shaky hypotheses.

But the episode of the *Parensae parumbrosae* is emblematic of what is happening in the most recent phases of parallel botany. As we have seen in our brief review, research is going on in many different directions, and though we do not yet have the comfort of clearly defined principles and the support of solid structures, what is emerging is a "style" of method and research that enables us to predict the general outlines of the new scientific discipline.

The circumstances of each new find enlarge our experience, and thereby increase the chances of further revelations. Special techniques are at last permitting us to transport plants which only a few years ago would seem to have been relegated forever to some dark and secret place of exile. In laboratories throughout the world, plants that have been, as it were, suspended for millennia between life and death now await the explanation of the mysteries of their existence.

The sudden questioning of things that have always and in every way conditioned our sensory and mental behavior demands a spirit of invention, an originality of method, a freedom of interpretation normally suffocated by the enormous weight of accepted ideas inherited as a result of our traditional scientific education. Thus it is that a growing number of young scientists, in spite of opposition from the establishment, are refusing to undertake research of which the results are a *fait accompli,* and instead are committing themselves with feverish enthusiasm to the exploration of an unknown world rich in exciting possibilities.

In spite of the warnings of good sense and personal gain, these men have dared to discard from their cultural and scientific baggage all those officially consecrated ideas they worked so hard to acquire, and have shown themselves willing to start again, to invent methods capable of penetrating the mysteries of a Nature whose laws are hidden in some remote and unknown country of our imagination.

It is reported of the Swedish philosopher Erud Kronengaard that he once said to a friend: "There are two kinds of men, those who are capable of wonder and those who are not. I hope to God that it is the first who will forge our destiny." A statement which strangely but clearly echoes the question put by Heraclitus to Theaetetus, a question to which the scientists now exploring the "other" reality beyond the hedge have already given a resoundingly emphatic answer.

NOTES

PART I: INTRODUCTION

General Introduction

1. Franco Russoli, *Una botanica inquietante* (Il Milione, Milan, 1972).

2. A vegetable-lamb is one of the illustrations in the *Voiage and Travayle of Sir Jhon Mandeville, Knight,* published in London in 1568. This animal-vegetable was also described by Parkinson in his *Theatrum Botanicum* in 1640, while a century later Erasmus Darwin mentioned it in his *Loves of the Plants.* It is known as Tartarian, Scytos or Vegetable-lamb, but most commonly as Tartarus or *Barometz.* First found in the Talmud, it appears in Europe during the Middle Ages (1330) in the writing of Odorico di Pordenone, in the form of a lamb fixed to a tree trunk which fed off the grass around the tree. In his *Histoire admirable des plantes et des herbes* (Paris, 1605), Claude Duret describes *Barometz* as a lamb whose wool is exceptional for its softness and beauty.

3. Romeo Tassinelli, *Scienze al traguardo,* various authors (Laguna, Venezia, 1972), pp. 105–22.

4. Remo Gavazzi, "Una rivoluzione vegetale" (*Corriere di Verona,* April 12, 1970).

5. In Hindu mythology Hanuman is a central figure of the Ramayana. Son of a nymph and the god of the winds, Hanuman, aided by an army of monkeys, helped Rama to save his wife Sita from the demon Ravana by taking boulders from the Himalayas to build a bridge between India and Ceylon.

6. *Antola enigmatica* is a plant that grows in the Indian state of Orissa as well as in certain parts of Central and South America. Its leaves form minute cups which fill with dew which then passes drop by drop into the cell-tissue of the plant.

7. Jacques Dulieu, *Un autre jardin* (Éditions La Nuit, Paris, 1973).

8. Max Spinder, "Wachsen und Zeitbegriff" (*Biologische Forschungen,* Basel, April 1968).

9. The Zendon Games are the Go championships played every year in Tokyo on the occasion of the festivities known as "Okiri." Go is the Japanese national game, in some ways resembling chess.

Origins

1. A small mining town in West Germany. The castle commands a view of the picturesque valley of Tiefenau. Apart from its important collection of fossils, Hochstadt is famous for the little Hochstadterhof, a hotel where paleontologists from all over the world have left signed photographs in the Fossilien Stube.

2. Piero Leonardi, "La Vita si evolve," in *L'Uomo, l'Universo, la Scienza* (Edindustria Editoriale, Rome [no date]).

3. Aktur (2680–2615 B.C.) was Emperor of the Anamids. In spite of his insane sadism he reigned for nearly thirty years. He was eventually killed by one of his 120 wives in a harem conspiracy. The episode was immortalized many centuries later by the Persian poet Hayem Ajaf Nazirim.

Morphology

1. C. H. Waddington, "The Character of Biological Form," in *Aspects of Form: A Symposium on Form in Nature and in Art*, edited by Lancelot Law Whyte (Pellegrini and Cudahy, New York, 1951).

2. Adolf Portmann uses the term "self-presentation" (in *Le forme viventi*, Adelphi, Milan, 1969) to mean the sum total of the external characteristics of living organisms and their "coming to light." He explains: "It is composed not only of the optical, acoustic and olfactory characteristics of the individual in a state of repose, but also of all those manifestations of himself in time and space which go beyond the functions of preservation, selection and immediate utility."

3. Wolfgang Keller, *Erscheinungslehre* (Institut für Geometrische Forschungen, Düsseldorf, 1970).

4. D'Arcy Wentworth Thompson, *On Growth and Form* (The Syndics of the Cambridge University Press, 1917).

5. Aldo Montù, *Natura e geometria* (Melocchi Editore, Milan, 1970).

6. Eric Hamilton, "Parallel Color," in *Biological Research, A Symposium* (Hartman and Coyle Ltd., London, 1969).

7. Marcello Vanni has been director of the famous Laboratory of Parallel Botany at Le Campora, Radda-in-Chianti, since 1969. He graduated from the University of Vienna, where he studied under Hermann von Kranenbogen. At Holtsville University in 1964 he gave a series of lectures on "Diagonality in Botany" which were the prelude to his present activities. He has recently been specializing in studies on the *Solea* and processes of transhabitation.

PART II: THE PLANTS

The Tirillus

1. An American poet of the school gravitating around the City Lights Bookshop in San Francisco. He has traveled widely in Central and South America collecting drug-bearing plants. He has published a collection of poems and stories (*Pot and Peyote*) and a short autobiographical novel (*Who? Me?*).

2. Roger Lamont-Paquit, French botanist, formerly lecturer in tropical botany at Lyons University. Since 1971 he has devoted himself entirely to the discovery and identification of parallel plants in South America.

The Tubolara

1. Note suppressed.

The Camporana

1. The number of leaves left on the *tialé* (36) and its multiples figure widely in the magic and mythology of ancient and primitive peoples. Thirty-six is the number of "Cosmic Solidarity," it is the "Grand Total" of the Chinese and the "Divine Year" of the Hindus. The numbers 36, 72 and 108 are favorites with secret societies, 36 being the number of the sky, 72 of the earth and 108 of man. There are 108 columns in the Temple of Ourga and 108 towers in Pnom Bakheng and Angkor. It is one of the symbolic numbers of Tantrism. According to Maspero, 72 and 108 divided by two give the astronomical coordinates of Loyang, the ancient Chinese Imperial capital.

2. Unlike the ritual candlestick, the seven "arms" of *C. menorea* are joined together to form one body, while at the top of each there is a small depression holding a paramimetic "seed" similar to those of the *Giraluna*.

3. A deadly poisonous plant similar to *Tahana fremens*. The Tuogoho fear its traitorous perfume, which if inhaled even briefly can cause death. They cal it the Flower of Huà. In their mythology Huà is a devil with a thousand heads, and in each head a different cruel thought.

The Protorbis

1. Pierre-Paul Dumasque, "La Protorbis de Katachek" (*Journal des Sciences Biologiques*, December 1969).

2. In official Hindu mythology, unlike what we find in this legend, Nandi is Shiva's bullock.

The Labirintiana

1. Cf. Eugène N. Marais, *L'Anima della formica bianca* (Adelphi, Milan, 1968).

The Artisia

1. Arthur Baldheim, *Many Adams* (Yale University Press, New Haven, 1957).

The Giraluna

1. Johannes Hydendorp, *De Maandbloem* (Nederlandse Uitgeversmaatschappij, Honingen, 1973).

2. The so-called *Giraluna* of Sommacampagna is a bronze which Giacomo Roselli has dated at the third century B.C. It was discovered while digging the foundations of a cooperative winery near Caselle di Sommacampagna (Verona). We do not know if this magnificent piece of sculpture is a copy or a reconstruction based on tales or old legends. The only other object found during the same excavations was a pot decorated with small round protuber-

ances the same size as the spherostills of the *Giraluna*. These occur at regular intervals around its rim.

3. Anselmo Geremia, "Un ponte fra le botaniche?" (*Quaderni Benenton*, January 1968). Geremia is Curator of the Parallel Botany Department of the Fabrizio Benenton Natural History Museum, Verona.

4. Born in Bierskonov in 1820 to a family of impoverished peasants, Oncharov died in Nizhni Novgorod in 1888. He wrote a number of ambitious but unsuccessful novels. "The Flower with the Golden Seeds" is one of the stories in *The Perfumed Sky of Biersk*, a book for children and the only one of his works to earn him some repute.

5. Harold Wittens, *Under the Wombasa Sky* (Simmons and Son, Liverpool, 1937).

6. In Thamed mythology Yakanan and Drapanias represent the anthromorphization of the *lingam* and the *yoni*. During the three days of the Hamadush the Yakanan on horseback chase the Drapanias, who flee from them swiftly on winged gazelles. Finally they copulate with them in the *Lahamna* position, holding the symbolic silver whips above their heads.

7. Marshall Norton, "Four American Photographers" (Hyford Gallery, London, 1965).

8. Albert Maessens, *An Adventure in Parallel Botany* (George Allen Thomas, London, 1972).

9. Johannes Hydendorp, *Parallel Ecology and Plant Behavior* (Van der Vos, Amsterdam, 1974).

The Solea

1. Max Spinder, *Die Solea—eine botanische Unentwicklung* (Univerlag, Hemmungen, 1972).

2. Antonio Guerrero, *Flora desconhecida do Rio* (Editorial Z, Rio de Janeiro, 1872).

3. John Foreman, *Flora South of the Border* (Henderson and Co., Boston, 1906).

4. *Soleares* (GLM, Paris, 1960).

5. Leo Lionni, not to be confused with the author of the present volume, is the pseudonym of Pieter Jacob Grossouwski, a name which the celebrated writer and illustrator abandoned early in his career in favor of one more easily pronounceable.

6. Randolph Reich, *Botanical Psychogenesis* (Harper & Row, New York, 1973).

7. Angel Pedro Maria Solinez, "Un sueño vegetal" (*Vida*, March 1973, Editorial de Mayo, Buenos Aires).

PART III: EPILOGUE

The Gift of Thaumas

1. For the Greeks Thaumas was the god of wonder. In the Platonic dialogue referred to by Altenhower, Socrates says: "Wonder is the emotion proper to the philosopher and philosophy begins in wonder. He was a wise genealogist who said that Iris, messenger of the heavens, was the child of Thaumas." (Jebb trans.)

A Note About the Author

Leo Lionni, a painter and writer—and now foremost parallel botanist—was born in Amsterdam in 1910. In 1934 he moved to Italy and in 1939 to the United States, where he worked as an artist and taught in various universities and art schools. For more than twelve years he was art director of *Fortune* magazine. In 1960 he returned to Italy and has since devoted himself to art, to writing children's books (over a dozen thus far, including the well-known *Little Blue and Little Yellow* and *Frederick*), and to exploring the world of parallel plants.

A Note on the Type

The text of this book was set on the Linotype in a new face called Primer, designed by Rudolph Ruzicka, who was earlier responsible for the design of Fairfield and Fairfield Medium, Linotype faces whose virtues have for some time been accorded wide recognition.

The complete range of sizes of Primer was first made available in 1954, although the pilot size of 12-point was ready as early as 1951. The design of the face makes general reference to Linotype Century—long a serviceable type, totally lacking in manner or frills of any kind—but brilliantly corrects its characterless quality.

Composed by American Book–Stratford Press, Brattleboro, Vermont. Printed and bound by Halliday Lithographers, West Hanover, Massachusetts.

Typography and binding design by Camilla Filancia.